DATE DUE

APR 19 '93			
APR 11 '95			
MAR 07 '96			
MAY 31 '96			
DEC 16 1996			

Hazardous Waste Minimization

Harry Freeman Editor

Risk Reduction Engineering Laboratory
U.S. Environmental Protection Agency

McGraw-Hill Publishing Company

New York   St. Louis   San Francisco   Auckland   Bogotá
Caracas   Hamburg   Lisbon   London   Madrid   Mexico
Milan   Montreal   New Delhi   Oklahoma City
Paris   San Juan   São Paulo   Singapore
Sydney   Tokyo   Toronto

Library of Congress Cataloging in Publication Data

Hazardous waste minimization/Harry Freeman, editor.
    p.   cm.
  Includes bibliographies and index.
  ISBN 0-07-022043-3
  1. Hazardous wastes—Management. 2. Waste minimization.
I. Freeman, Harry, date.
TD1050.W36H39 1990
363.72'877—dc20

89-35068
CIP

34567890 DOC/DOC 954321

ISBN 0-07-022043-3

The editors for this book were Robert W. Hauserman and Caroline Levine, the designer was Naomi Auerbach, and the production supervisor was Richard A. Ausburn. It was set in Century Schoolbook and was composed by the McGraw-Hill Publishing Company Professional & Reference Division composition unit.

Printed and bound by R. R. Donnelley and Sons Company.

Contents

Section 4. Case Studies of Successful Waste Minimization Programs

Contributors

W. B. Beck *E. I. duPont, Houston, Texas* (CHAP. 4)

Denny Beroiz *General Dynamics, Pomona, California* (CHAP. 15)

Robert J. Chabot *San Antonio Air Logistics Center, Kelly Air Force Base, Texas* (CHAP. 16)

Harry M. Freeman *Risk Reduction Engineering Laboratory, U. S. Environmental Protection Agency, Cincinnati, Ohio* (CHAP. 1)

Carl Fromm *Jacobs Engineering Group, Pasadena, California* (CHAP. 5)

Tim Garrett *Anniston Army Depot, Anniston, Alabama* (CHAP. 16)

Deborah Hanlon *Jacobs Engineering Group, Pasadena, California* (CHAP. 5)

Margaret Harris *Warner Robins Air Logistics Center, Robins Air Force Base, Georgia* (CHAP. 16)

Joseph Hoenscheid *Defense Logistics Agency, Cameron Station, Alexandria, Virginia* (CHAP. 16)

Gregory Hollod *Conoco, Inc., Houston, Texas* (CHAP. 4)

Gary E. Hunt *North Carolina Pollution Prevention Program, Raleigh, North Carolina* (CHAP. 3)

Joseph A. Kaminski *Office of the Deputy Assistant Secretary of Defense (Environment), Washington, D.C.* (CHAP. 16)

Donald W. Koepp *Ventura County Environmental Health Department, Ventura, California* (CHAP. 13)

James Lounsbury *Office of Solid Waste, U. S. Environmental Protection Agency, Washington, D. C.* (CHAP. 1)

Ronald T. McHugh *Office of Policy, Planning, and Evaluation, U. S. Environmental Protection Agency, Washington, D.C.* (CHAP. 6)

Tony Pollard *Anniston Army Depot, Anniston, Alabama* (CHAP. 16)

Linda Giannelli Pratt *Hazardous Materials Management Division, San Diego County Department of Health Services, San Diego, California* (CHAP. 14)

Scott M. Raymond *Raymond Environmental Compliance Services, Glenshaw, Pennsylvania* (CHAP. 9)

Steven Rice *BASF Corporation, Parsippany, New Jersey* (CHAP. 8)

Robert H. Salvesen *Robert H. Salvesen Associates, Tinton Falls, New Jersey* (CHAP. 19)

Roger N. Schecter *North Carolina Pollution Prevention Program, Raleigh, North Carolina* (CHAP. 12)

Susan Sherry *California Local Government Commission, Sacramento, California* (CHAP. 11)

Dale A. Sowell *Department of the Navy, Naval Sea Systems Command, Washington, D.C.* (CHAP. 16)

John Warren *Center for Economics Research, Research Triangle Institute, Research Triangle Park, North Carolina* (CHAP. 2)

David Wigglesworth *Waste Reduction Assistance Program (WRAP), Alaska Health Project, Anchorage, Alaska* (CHAPS. 7 and 10)

Ron Willenbrink *Ashland Oil, Inc., Ashland, Kentucky* (CHAP. 17)

Katy Wolf *Source Reduction Research Partnership, Los Angeles, California* (CHAP. 18)

Preface

As environmental regulations have become more stringent in industrial countries throughout the world, it is becoming increasingly obvious that ultimate answers lie in eliminating the generation of hazardous wastes altogether rather than just treating the wastes to insure environmentally acceptable disposal. The strategy to encourage such waste elimination has assumed various names, such as *low-* and *non-waste technologies, clean technologies, clean products, waste reduction,* and *pollution prevention.* The name which has assumed a high degree of popularity in the United States is *waste minimization,* the term used in the title of this book.

In compiling the materials for this book, which is designed to assist individuals in private companies and public agencies in identifying and pursuing options for waste minimization, I asked recognized professionals in the field for contributions. Their chapters provide insights from many useful perspectives, which I believe will be very useful in reducing or eliminating hazardous waste.

I encourage you to use their suggestions to bring about what we are all working for—reduced environmental problems caused by hazardous waste.

Harry Freeman

Acknowledgments

Many people made this book happen: the contributing authors, who not only provided well-written manuscripts, but provided them in a timely manner; my colleagues at the EPA and throughout the country, who responded so well to my requests for suggestions and criticisms; and finally, the editors at McGraw-Hill who were super. Thanks to all of you.

Waste Minimization Overview

Waste Minimization as a Waste Management Strategy in the United States

Harry M. Freeman

Risk Reduction Engineering Laboratory
U.S. Environmental Protection Agency
Cincinnati, Ohio

James Lounsbury

Office of Solid Waste
U.S. Environmental Protection Agency
Washington, D.C.

The Current Status of Waste Minimization in the United States

The term *waste minimization* is heard increasingly at meetings and conferences by individuals working in the field of hazardous waste management. Waste minimization means the reduction, to the extent feasible, of hazardous waste that is generated prior to treatment, storage, or disposal of the waste. The United States Environmental Protection Agency (U.S. EPA) encourages the minimization of all wastes that pose risks to human health and the environment. Waste minimization techniques focus on source reduction or recycling activities that reduce either the volume or the toxicity of hazardous waste generated. Unlike many approaches to waste treatment, waste minimization can be practiced at several stages in most industrial processes. Like all innovative solutions to waste management problems, waste minimization requires careful planning, creative problem solving, changes in attitude, sometimes capital investment, and, most important, a real commitment.

The payoffs for this commitment, however, can be great. Waste minimization can save money—often substantial amounts—through more efficient use of valuable resources and reduced waste treatment and disposal costs. Waste minimization also can reduce a generator's hazardous-waste-related financial liabilities; the less waste generated, the lower the potential for negative environmental effects. Finally, taking the initiative to reduce hazardous waste is good policy. Polls in the United States show that reducing toxic chemical risk is the public's primary environmental concern. Waste minimization can pay off tangibly when local residents are confident that industry is making every effort to handle its wastes responsibly.

Various individuals and agencies use terms other than *waste minimization* to denote the same approach. The U.S. EPA also uses the term *pollution prevention* to refer to the same concept applied to all releases of substances to the air, land, and water. A listing of these somewhat synonymous terms is shown in Table 1.1.

Figure 1.1 illustrates waste minimization techniques. Some industry-specific examples are shown in Table 1.2. It should be noted that the term does *not* include such processes as incineration, stabilization, or storage, although these processes are clearly environmentally preferable to land disposal of untreated wastes. Such processes are just lower on the waste management hierarchy shown in Table 1.3. The idea underlying waste minimization, the top two terms on the hierarchy, is that it makes far more sense for a generator not to produce waste rather than to develop extensive treatment schemes to insure that the wastestream poses no threat to the quality of the environment.

The U.S. Congress specifically stated in the Hazardous and Solid Waste Amendments of 1984 to the Resource Conservation and Recovery Act (RCRA):

TABLE 1.1 Somewhat Synonymous Terms for Waste Minimization

Waste minimization

Waste reduction

Clean technologies

Pollution prevention

Environmental technologies

Low- and nonwaste technologies

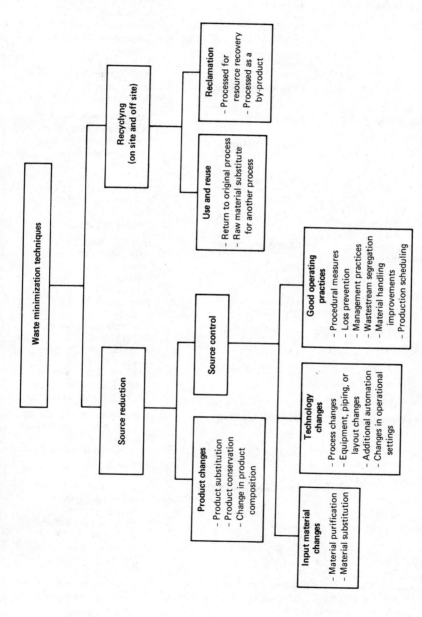

Figure 1.1 Waste minimization techniques.

TABLE 1.2   Examples of Waste Minimization Techniques

Technique category	Industry	Technique
Inventory management	Textiles	Review all chemical purchases.
	Furniture	Purchase only exact amounts of coating required.
	Diesel engines	Screen all products purchased.
	Organic chemicals	Review new products before production.
Material change	Printing	Substitute water-based for solvent-based ink.
	Office furniture	Use water-based paints in place of solvent-based paints.
	Aerospace	Replace cyanide cadmium-plating bath with noncyanide bath.
	Ink manufacture	Remove cadmium pigment from products.
Production process modifications	Chemical reaction	Optimize reaction variables and reactor design. Optimize reactant addition method. Eliminate use of toxic catalysts.
	Surface coating	Use airless air-assisted spray guns. Use electrostatic spray coating system. Control coating viscosity with heat units. User high-solids coatings.
In-plant recycling	Metal fabricators	Recover synthetic cutting fluids using a centrifuge system.
	Paint fabricating	Use distillation unit to recover cleaning solvents.
	Printed circuit boards	Use an electrolytic recovering system to recover copper and tin/lead from process wastewater.
	Power tools	Recover alkaline degreasing baths using ultrafiltration.

SOURCE: *Journal of the Air Pollution Control Association*, Vol. 38, No. 1, January 1988.

TABLE 1.3   **Waste Management Hierarchy**

Source reduction	The reduction or elimination of hazardous waste at the source, usually within a process. Source reduction measures include process modifications, feedstock substitutions, improvements in feedstock purity, housekeeping and management practice changes, increases in the efficiency of equipment, and recycling within a process.
Recycling	The use or reuse of hazardous waste as an effective substitute for a commercial product or as an ingredient or feedstock in an industrial process. It includes the reclamation of useful constituent fractions within a waste material or the removal of contaminants from a waste to allow it to be reused.
Treatment	Any method, technique, or process which changes the physical, chemical, or biological character of any hazardous waste so as to neutralize such waste, to recover energy or material resources from the waste, or to render such waste nonhazardous, less hazardous, safer to manage, amenable for recovery, amenable for storage, or reduced in volume.
Disposal	The discharge, deposit, injection, dumping, spilling, leaking, or placing of hazardous waste into or on any land or water so that such waste or any constituents may enter the air or be discharged into any waters, including groundwater.

The Congress hereby declares it to be the national policy of the United States that, wherever feasible, the generation of hazardous waste is to be reduced or eliminated as expeditiously as possible. Waste that is nevertheless generated should be treated, stored, or disposed of so as to minimize the present and future threat to human health and the environment.

Other organizations, including the U.S. Congress Office of Technology Assessment, the National Academy of Sciences, the EPA's Science Advisory Board, the Environmental Defense Fund, and the Natural Resources Defense Council, have issued strong statements in support of programs to encourage the development and adoption of waste minimization strategies. The EPA's position on the subject was succinctly detailed in the report it gave to Congress in 1986:

EPA still has much to learn about waste minimization and recognizes that the cooperation of private and public waste generators will be invaluable as it moves toward the development of sound long-term policy. It also believes, however, that the incentives and trends within the hazardous waste management system are unmistakable, and that the program presented here comprises the most positive and constructive steps that can be taken at this time. Aggressive action in favor of waste minimization is clearly needed, but a major new regulatory program—at least for the present—does not seem desirable or feasible.

Incentives for waste minimization are already strong, so EPA must

capitalize on them. Most lacking is access by generators to the information that will demonstrate the economic benefits of waste minimization
to industry, overcome logical problems, and help develop creative new approaches. This can be provided by a strong technical assistance and information transfer effort, which can achieve through voluntary means
what would be inefficient and possibly counterproductive to attempt
through regulation. Unfortunately, non-regulatory programs have often
failed at EPA for lack of statutory or regulatory deadlines and institutional advocacy. For such a program to work, it must be given strong organizational support within the Agency. EPA is willing to make this
commitment, and seeks support from Congress to ensure its success.

In the United States, the Hazardous and Solid Waste Amendments
of 1984 requires every hazardous waste generator to report on activities undertaken to reduce the volume and toxicity of wastestreams
and to certify that he or she has a program in place to minimize waste
generation. The certification statement is

> If I am a large-quantity generator, I certify that I have a program in
> place to reduce the volume and toxicity of waste generated to the degree
> I have determined to be economically practicable and that I have selected
> the practicable method of treatment, storage, or disposal currently avail-
> able to me which minimizes the present and future threat to human
> health and the environment, or, if I am a small-quantity generator, I
> have made a good-faith effort to minimize my waste generation and se-
> lect the best waste management.

The EPA's draft guidance for what constitutes an effective waste minimization program is reprinted at the end of this chapter. This guidance recognizes that programs may vary from plant to plant. What
works for a large facility may make no sense for a small facility, or
vice versa.

Although the EPA does require that generators have a waste minimization program, there are no current federal regulations that require specific waste minimization accomplishments or quotas on the
part of generators. This situation is based in part on the EPA's findings in its 1986 report to Congress on hazardous waste minimization.
The EPA report to Congress, entitled *Minimization of Hazardous
Waste* (EPA 530-SW-86-0330), was submitted by the EPA to the Congress in October 1986. In this report the agency explored many technical, economic, and policy issues pertinent to reducing waste generation and concluded that it would be counterproductive for EPA to
establish a mandatory program for waste minimization.

The conclusion was based on three key factors. First, mandatory
programs would second-guess industry's production decisions, quite
possibly leading to counterproductive results. Second, mandatory pro-

grams would be difficult and expensive to design and administer. Third, generators already face strong economic incentives to reduce their wastes. A regulatory program would take time to develop, and many industries might postpone any action until mandatory requirements were spelled out.

The 1986 report to Congress stressed that the most constructive role government can assume is to promote voluntary waste minimization by providing information, technology transfer, and assistance to waste generators. The report further concluded that even though a mandatory regulation program is not needed, "aggressive action in favor of waste minimization is clearly needed." A three-point strategy was outlined to guide the EPA in preparing an aggressive program:

1. *Information gathering:* Detailed data on industry's response to the land disposal restrictions program and other existing waste minimization incentives would be gathered in order to make a final determination on the desirability and feasibility of performance standards and required management practices.

2. *Core technical assistance program:* During the interval when the new provisions of the Hazardous and Solid Waste Amendments of 1984 are taking effect, EPA would launch a strong technical assistance and information transfer program through the states to promote voluntary waste minimization in industry, government, and the non-profit sectors of the economy. It would also work with federal agencies to encourage procurement practices that promote the use of recycled and reclaimed materials.

3. *Long-term options:* Based on an analysis of the new data gathered under the first part of this three-point strategy, performance standards and other mandatory requirements could be imposed, if necessary, once the Hazardous and Solid Waste Amendments of 1984 have taken full effect and their various impacts on waste generation have been assessed.

The EPA has promised to submit another report to Congress in 1990 that will revisit the question of mandatory regulations. This report will incorporate the results of the agency's current activities to further its state of knowledge on the subject.

The Future of Waste Minimization in the United States

Waste minimization is an essential element of the nation's immediate and long-term strategy to manage hazardous waste. Land disposal

TABLE 1.4   Land Ban Restrictions Timetable

Dioxins and solvents	November 8, 1986
California List (metals and cyanides, corrosives, halogenated organics)	July 8, 1987
First third of remaining hazardous wastes	August 8, 1988
Second third of remaining hazardous wastes	June 8, 1989
All hazardous wastes	May 8, 1990

will continue to play a role, but that role is diminishing. EPA's land disposal restrictions program (see Table 1.4) will ban the land disposal of many untreated hazardous wastes, forcing generators to explore other options. Treatment technologies can assume much of the waste management burden from land disposal, but treatment is expensive, and, at least in the near term, capacity is limited. EPA's strategy to minimize the generation of hazardous waste will help reduce or eliminate regulated wastes that are now managed by treatment or land disposal as well as other wastes that pose risks to human health and the environment.

Waste minimization clearly provides opportunities to deal more efficiently and effectively with wastes that are hazardous to human health and to the environment. These opportunities are unique in that they provide immediate financial rewards to industry, increased waste management flexibility to generators, and reduced pressures on the nation's existing treatment and land disposal capacity. Now is the time to investigate and take practical steps toward waste minimization, before major commitments are made for treatment and disposal options. Over the longer term, the benefits of source reduction and recycling will be key incentives for generators to integrate waste minimization techniques into their overall hazardous waste management programs.

GUIDANCE TO HAZARDOUS WASTE GENERATORS ON THE ELEMENTS OF A WASTE MINIMIZATION PROGRAM (FEDERAL REGISTER/JUNE 12, 1989)

I. PURPOSE

The purpose of today's notice is to provide non-binding guidance to generators of regulated hazardous wastes on what constitutes a "program in place" to comply with the certification requirements of Sections 3002(b) and 3005(h) of the Solid Waste Disposal Act, as amended by the Resource Conservation and Recovery Act (RCRA) and the Hazardous and Solid Waste Amendments of 1984 (HSWA). Such certifications require generators to implement programs to reduce the volume and toxicity of hazardous wastes generated to the extent economically practicable. This guidance is intended to fulfill a commitment made by EPA in its 1986 report to Congress entitled, *Minimization of Hazardous Waste*.[1]

II. BACKGROUND

With the passage of HSWA, Congress established a national policy declaring the importance of reducing or eliminating the generation of hazardous waste. Specifically, Section 1003(b) states:

> The Congress hereby declares it to be a national policy of the United States that, wherever feasible, the generation of hazardous waste is to be reduced or eliminated as expeditiously as possible. Waste that is nevertheless generated should be treated, stored, or disposed of so as to minimize present and future threat to human health and the environment.

In this declaration, Congress established a clear priority for reducing or eliminating the generation of hazardous wastes (a concept referred to as waste minimization) over managing wastes that were "nevertheless" generated.

EPA believes that hazardous waste minimization means the reduction, to the extent feasible, of hazardous waste that is generated prior to treatment, storage, or disposal of the waste. It is defined as any source reduction or recycling activity that results in either: (1) reduction of total volume of hazardous waste; (2) reduction of toxicity of hazardous waste; or (3) both, as long as that reduction is consistent with the general goal of minimizing present and future threats to human health and the environment.[2]

Waste minimization can result in significant benefits for industry. EPA believes an effective waste minimization program will contribute to

1. minimizing quantities of regulated hazardous waste generated, thereby reducing waste management and compliance costs;
2. improving product yields;
3. reducing or eliminating inventories and releases of "hazardous chemicals" reportable under Title III of the Superfund Amendments and Reauthorization Act; and/or
4. lowering Superfund, corrective action and toxic tort liabilities.

Besides establishing the national policy, Congress also enacted several provisions in HSWA for implementing hazardous waste minimization. These included a generator certification on hazardous waste manifests

---

[1] 51 Fed. Reg. 44683 (12/11/86), Notice of Availability of the report to Congress.

[2] Hazardous waste minimization involves volume or toxicity reduction through either a source reduction or recycling technique and results in the reduction of risks to human health and the environment. The transfer of hazardous constituents from one environmental medium to another does not constitute waste minimization. Neither would concentration conducted solely for reducing volume unless, for example, concentration of the waste allowed for recovery of useful constituents prior to treatment and disposal. Likewise, dilution as a means of toxicity reduction would not be considered waste minimization unless later recycling steps were involved.

and permits for treatment, storage, or disposal of hazardous waste. RCRA §3002(b). These certifications (effective September 1, 1985) require generators to certify two conditions: that (1) the generator of the hazardous waste has a program in place to reduce the volume or quantity and toxicity of such waste to the degree determined by the generator to be economically practicable; and (2) the proposed method of treatment, storage or disposal is that practicable method currently available to the generator which minimizes the present and future threat to human health and the environment.

In addition, Congress also added a new provision in 1984 that requires hazardous waste generators to identify in their biennial reports to EPA (or the State): (1) the efforts undertaken during the year to reduce the volume and toxicity of waste generated; and (2) the changes in volume and toxicity actually achieved in comparison with previous years, to the extent such information is available prior to 1984 (RCRA §3002(a)(6)).

Today's notice provides non-binding guidance to hazardous waste generators in response to the certification requirements in HSWA. Specifically, it addresses the first of the certification conditions that states that, "the generator of the hazardous waste has a program in place to reduce the volume or quantity and toxicity of such waste to the degree determined to be economically practicable."

EPA is not, however, providing guidance on the determination of the phrase "economically practicable." As Congress indicated in its accompanying report to HSWA[3], the term "economically practicable" is to be defined and determined by the generator and is not subject to subsequent re-evaluation by EPA. The generator of the hazardous waste, for purposes of this certification, has the flexibility to determine what is economically practicable for the generator's circumstances. Whether this determination is made for all of its operations or on a site-specific basis is for the generator to decide.

EPA has received numerous inquiries on what constitutes a waste minimization program. In today's notice EPA is providing draft guidance to hazardous waste generators on what the Agency believes are the basic elements of a waste minimization program.

III. ELEMENTS OF A WASTE MINIMIZATION PROGRAM

EPA believes that today's guidance may provide direction to large quantity and small quantity generators in fulfilling their manifest certification requirement. Small quantity generators, while not subject to the same "program in place" certification requirement as large quantity generators, have to certify that they have "made a good faith effort to minimize" their waste generation.

The elements discussed here reflect the results of agency analyses conducted over the last several years and extensive interaction with private

---

[3] S.Rep.No. 98-284, 98th Cong., 1st Sess. (1983)

and public sector waste minimization program managers. EPA believes that an effective waste minimization program should include each of the general elements discussed below, although EPA realizes that some of these elements may be implemented in different ways depending on the preferences of individual firms.

A. *Top management support.* Top management should ensure that waste minimization is a company-wide effort. There are many ways to accomplish this goal. Some of the methods described below may be suitable for some firms and not others. However, some combination of these techniques should be used by every firm to demonstrate top management support.

- Make waste minimization a company policy. Put this policy in writing and distribute it to all departments. Make it each person's responsibility to identify opportunities for minimizing waste. Reinforce the policy in day to day operations, at meetings and other company functions.
- Set specific goals for reducing the volume or toxicity of wastestreams.
- Commit to implementing recommendations identified through assessments, evaluations or other means.
- Designate a waste minimization coordinator at each facility to ensure effective implementation of the program.
- Publicize success stories. It will trigger additional ideas.
- Reward employees that identify cost-effective waste minimization opportunities.
- Train employees on aspects of waste minimization that relate to their job. Include all departments, such as those in product design, capital planning, production operations, and maintenance.

B. *Characterization of waste generation.* Maintain a waste accounting system to track the types, amounts and hazardous constituents of wastes and the dates they are generated.

C. *Periodic waste minimization assessments.* Track materials that eventually wind up as waste, from the loading dock to the point at which they become a waste.

- Identify opportunities at all points in a process where materials can be prevented from becoming a waste (for example, by using less material, recycling materials in the process, finding substitutes, or making equipment changes.) Individual processes or facilities should be reviewed periodically. Larger companies may find it useful to establish a team of independent experts.
- Determine the true costs of the waste. Calculate the costs of the materials found in the waste stream based on the purchase price of those materials. Calculate the cost of managing the wastes that are generated, including costs for personnel, record keeping, transportation, liability insurance, pollution control equipment, treatment and disposal and others.

D. *A cost allocation system.* Departments and managers should be charged "fully-loaded" waste management costs for the wastes they generate, factoring in liability, compliance and oversight costs.

E. *Encourage technology transfer.* Seek or exchange technical information on waste minimization from other parts of your company, from other firms, trade associations, State and university technical assistance programs or professional consultants. Many techniques have been evaluated and documented that may be useful in your facility.

F. *Program evaluation.* Conduct a periodic review of program effectiveness. Use these reviews to provide feedback and identify potential areas for improvement.

Although waste minimization practices have demonstrated their usefulness and benefits to those generators that have implemented such programs, many others still have not practiced waste minimization. Today's guidance on effective waste minimization practices may help encourage regulated entities to investigate waste minimization alternatives, implement new programs, or upgrade existing programs. Although the approaches described above are directed toward minimizing hazardous solid waste, they are equally valid for design of multimedia source reduction and recycling programs.

Potential for Waste Reduction

John L. Warren

Center for Economics Research
Research Triangle Institute
Research Triangle Park, North Carolina 27709

The following factors have a significant influence on the potential for waste reduction:

- What wastes are of interest
- How we define waste reduction
- What factors affect the quantity of waste considered available for waste reduction

Sections 2.2 through 2.4 review these factors briefly in the context of estimating the potential for waste reduction. Section 2.5 provides estimates of the quantities of wastes generated; estimates of the potential for waste reduction are only meaningful to the extent that we can understand the amount of waste actually being generated. Table 2.4 in Sec. 2.5 summarizes the estimates of waste reduction potential from a variety of government and nongovernment sources. Section 2.6 provides a list of reference sources for the information in the tables and text of this chapter.

2.1 Hazardous Wastes of Interest

Hazardous wastes can be defined in a variety of ways. We will focus on wastes considered hazardous under the Resource Conservation and Recovery Act (RCRA). A waste considered to be hazardous under the RCRA must be either a solid waste, as defined under RCRA, that is a listed waste from a specific production process, or a solid waste, as defined under RCRA, that exhibits at least one of the four characteristics of a hazardous waste—corrosivity, reactivity, ignitability, or toxicity. Specific legal definitions of what constitutes a hazardous waste can be found at Section 40 CFR 261.3 of the RCRA. Some wastes do not fall under the RCRA definitions, yet they exhibit characteristics and behavior that could be associated with a hazardous waste. Such wastes might include the following:

- Wastes that are managed in units exempt from RCRA regulation and permitting, even though they are still hazardous (e.g., hazardous waste managed in tanks regulated under the Clean Water Act)

- Wastes that are hazardous only under state, but not federal, regulation (e.g., waste oils or certain wastes with pH levels between 2 and 4)

- Wastes that have been excluded due to Congressional action (e.g., selected mining and associated processing wastes)

Any estimate of the potential for waste reduction should carefully define the universe of wastes included in the estimate. Otherwise, the estimates will not provide useful benchmarks for measuring waste reduction progress. Minor changes in the words of a regulation can significantly change the number of tons of wastes that are considered hazardous and thus subject to waste reduction.

2.2 What Is Waste Minimization?

There have been numerous legal, regulatory, and scientific arguments as to what constitutes waste minimization. EPA has adopted a hierarchy of preferred waste management protocols as follows, from most preferred to least preferred: source reduction, recycling, waste reduction after generation, treatment, and disposal. Waste minimization is any activity considered to be source reduction or recycling.

2.3   Factors Affecting Estimates of Waste Reduction Potential

The Environmental Defense Fund (EDF), in studying the potential for waste reduction, found that several types of data are needed to estimate accurately the potential for waste reduction:

- Current wastestreams (volumes and characteristics)
- Future changes in waste volumes and characteristics
- Current waste management practices
- Current and anticipated federal and state regulations

Unfortunately, all these data were often not available when estimates of waste reduction potential were made. Consequently, estimates have to be evaluated with the understanding that they were made using incomplete data. Major changes in any of these data can result in incorrect reduction estimates. For example, if a corrosive hazardous waste were defined as having a pH $\leq$ 3 (instead of pH $\leq$ 2), significant volumes of wastewater that are not currently labeled hazardous would move into the hazardous category. This would then increase the volumes of hazardous wastes available for and requiring waste reduction.

The U.S. Congress Office of Technology Assessment (OTA) has completed several studies on the need for reducing the generation of hazardous wastes. OTA noted that several factors affect our ability to determine waste reduction potential:

- We lack substantive information.
- It is difficult to determine waste generated per-unit output values.
- Firms may not collect data needed to make calculations.
- Industrial activity, product mix, and environmental regulatory requirements change over time and affect the quantity and characteristics of wastes generated.
- Waste volume may be reduced but the hazard level of the waste may be the same or even greater; dilution disperses the same amount of waste but in larger volumes.

OTA concludes that:

> The experience with the RCRA waste generation data system shows that it would be remarkably difficult for government to gather and analyze accurate and timely data from a very large number of companies and for an even larger number of processes and waste streams. Even reliable data would not reveal what actions, out of a broad range of possibilities, a specific industrial operation might reasonably undertake to reduce

waste, because so much of what is technically and economically feasible is site-specific.[1]

2.4 Estimates of Hazardous Wastes Generated

In order to estimate the potential for waste reduction, we need to review the estimates of the volume of hazardous waste generated. These wastes can be classified into two broad categories:

1. Wastes generated on an annual or continuous basis. These include wastes from typical manufacturing, production, and service facilities.

2. Wastes generated on a sporadic or noncontinuous basis. These include primarily wastes from cleanups at Superfund and similar sites and from corrective action taken at facilities that have RCRA permits.

Table 2.1 summarizes estimates of the quantity of hazardous waste generated on a continuous basis in the United States. These wide

TABLE 2.1 Estimates of Quantity of Hazardous Wastes Generated

Hazardous wastes generated	Amount (MMT)	Year estimated	Comments
1986 National Screening Survey—EPA[2]	272 (in units regulated under RCRA) 313 (in units exempt from RCRA)	1985	Census of TSDR* facilities
Chemical Manufacturers Association (CMA)[3]	247 220.5	1984 1986	Chemical industries survey only
1981 National Survey—EPA[4]	247 (135–402)†	1981	Sample survey
Office of Technology Assessment (OTA)[1]	255–275†	1981	Data compilation from states
Congressional Budget Office (CBO)[5]	266 (223–308)†	1983	Model-based using disaggregated industry data

* TSDR = Treatment, Storage, Disposal, and Recycling
† Estimated range
Note: Superior numbers refer to sources in Reference list at end of chapter.

ranges are a reflection of the factors discussed in Sec. 2.3. The most significant factor affecting estimates of the total quantity of hazardous waste is how the survey or estimate handled hazardous wastewater managed in exempt units (such as tanks) that discharge to a publicly owned treatment works (POTW) or under a NPDES (National Pollutant Discharge Elimination System) permit. The data from the latest year for which data are available, 1986, suggests that the largest quantities of wastes are managed in these exempt units. Data for these units have not been collected in the past. However, EPA has several large data-collection efforts that will provide these data in 1989.

Operating practices and conditions at existing and former facilities used for managing hazardous wastes have and do frequently result in releases to air, soil, surface water, and groundwater. Currently, these problems are addressed by two major environmental programs: SARA (Superfund) and RCRA (as amended). RCRA focuses on existing, operating sites covered by RCRA permits. The RCRA program works to get facility owners and operators to implement required corrective actions to clean up contamination caused by operating hazardous waste facilities at the site. In contrast, under Superfund the focus is on abandoned and/or inactive sites where the facility's owners or operators are not known or are unwilling or financially incapable of paying to clean up any contamination.

Wastes generated by RCRA corrective action and Superfund cleanups are not generated continuously but may have a significant impact on the volumes of waste that the nation needs to manage. The volume of hazardous waste generated by a site is a function of technical, financial, health, and political decisions. Thus, there can be waste reduction even from a cleanup if waste reduction is incorporated into the decision-making process for that site. The U.S. General Accounting Office (GAO) estimates that the numbers and types of sites shown in Table 2.2 may exist and place demands on any system for managing hazardous wastes.

Table 2.3 shows the estimated number of facilities that may require corrective action at sites that are subject to RCRA permitting requirements.

2.5   Potential for Waste Reduction

OTA has noted that the degree of unrealized waste reduction potential is part of the waste minimization problem. Consequently, it is useful to review the range of estimates of the potential for reduction even though no two studies or estimates were made in the same way. OTA

TABLE 2.2 Estimates of Hazardous Wastes Generated on a Noncontinuous Basis

	Number of sites	
Type of site	1985	1987
RCRA waste facilities		
Nonhazardous (Subtitle D)	109,000–127,000	70,419–261,930
Hazardous (Subtitle C)	605	818
Mining waste sites	9770–63,770	22,339
Underground storage tanks (nonpetroleum)	11,250–187,500	10,820
Pesticide-contaminated sites	*	3920
Federal facility sites		
Civilian	*	1882
Defense	*	3526
Radioactive releases	*	300
Underground injection wells	*	13,839–117,368
Town gas facilities	*	1502
Wood-preserving plants	*	975
Total	130,625–378,875	130,340–425,380

*Data not available.

NOTE: This chart shows the estimates of sites that can potentially result in the generation of hazardous wastes that have to be treated and disposed of but which could also be conceivably amenable to waste reduction measures. These sites will add an additional burden to the nation's system for managing hazardous wastes.

SOURCE: U.S. General Accounting Office, December 1987, GAO/RCED-88-44, *Identification of Hazardous Waste Sites*, Table 2.1.

TABLE 2.3 Projected Number of RCRA Facilities That May Require Corrective Action and Result in the Generation of Hazardous Wastes

Projected no. of facilities	Operating land disposal	Closing land disposal	Operating & closing including trmnt. & storage	Total facilities
In universe (total)	393	1095	3338	4826
Requiring RFIs*	303	776	1869	2938
Requiring corrective action	243	652	1589	2484

* remedial investigation/feasibility studies

NOTE: EPA has estimated that (1) each of these facilities has an average of 6.6 nonregulated units and (2) between 27 percent and 54 percent of these units have leaked and may require cleanup.

SOURCE: U.S. General Accounting Office, December 1987, GAO/RCED-88-48, *Corrective Action Cleanups*, Table II.1

has listed a variety of reasons for which estimates of waste reduction potential have uncertainty:

- There are too many industrial processes and wastes—certainly tens of thousands—to examine each in detail.

- Waste generation and reduction are plant- and process-specific, but the limited waste generation data available are aggregated over many processes and usually over a diversity of plants and companies.

- It is not known how much waste in all (not just RCRA waste) was and is now being generated; therefore, reduction cannot be documented.

- There is no base year for all data.

- It is difficult to predict what changes in production technology and products will occur over a broad range of industry for reasons unrelated to waste, and such changes can substantially change the nature or quantity of waste or both.

- Considerable amounts of wastes (particularly as regulated under the Clean Water and Clean Air Acts) are legally sanctioned, and continued implementation of environmental programs will create more waste (e.g., pretreatment standards under the Clean Water Act increase the generation of solid hazardous wastes).

- Many regulatory, enforcemental, and judicial actions that affect the economic feasibility of and perceived need for waste reduction may occur.[1]

Table 2.4 summarizes the range of estimates of the potential for hazardous waste reduction—from 11 percent to almost 75 percent. This wide range reflects the associated differences in estimating the potentials and the universe of wastes included in the estimates. Often estimates are quoted out of context or without a full understanding of the factors used to make the estimates. Readers are urged to review the references for this chapter (given in Sec. 2.6) to gain a fuller appreciation for the potential for waste reduction.

TABLE 2.4   Estimates of the Potential for Waste Reduction

Source	Year estimate made	Estimated reduction	Comments
National Research Council (committee study)[7]	1985	Unable to make estimate	Not enough data available
Ventura County, California (survey of 75 firms)[8]	1985–1986	30–40%	Ventura County only; not national in scope; focus on reduction in volumes sent to landfills; often misquoted
U.S. Environmental Protection Agency (1986 Report to Congress)[9,10,11]	1986	18–33%	National in scope; aggregated; potential over a 25-year period; source controls only
3M Company (company experiences)[1]	1985	50% Additional 30%	Over past 10 years Over next 5 years
Office of Technology Assessment (OTA) Industry Survey (99 respondents)[1]	1985	50 to 75% possible 25 to 50% possible < 25% possible	11% of respondents 25% of respondents 59% of respondents
OTA opinion (study conclusions)[1,6]	1985	Substantial opportunities	No specific data available
Congressional Budget Office (CBO)[12]	1985	14% < 1983 amount	Based on model for 24 waste types generated in 1990 by 70 industries
Specific firms[1]	1986	Rohm & Haas—10% (1984 to 1985) Exxon Chem.—10% (1984 to 1985) Olin—34% (1981 to 1985) Dupont—35–50% 3M—50% (1975 to 1985) IBM—17% (1984 to 1985) Hewlett Packard —16% (1983 to 1984)	Adjusted for production changes

NOTE: Superior numbers refer to sources in Reference list at end of chapter.

Source	Year estimate made	Estimated reduction	Comments
INFORM[13]	1985	Up to 90%	Based on survey of 29 organic chemical plants
TVA region[14]	1984	11% for incinerable wastes 33% for landfilled wastes	Changes from 1984 to 2000
Minnesota[1]	1985	47% reduction in 2000 compared with 1984 volumes	RCRA wastes only

NOTE: Superior numbers refer to sources in Reference list at end of chapter.

2.6   REFERENCES

1. Office of Technology Assessment, *Serious Reduction of Hazardous Waste: For Pollution Prevention and Industrial Efficiency*, U.S. Government Printing Office, September 1986.
2. U.S. Environmental Protection Agency, "1986 National Screening Survey of Hazardous Waste Treatment, Storage, Disposal, and Recycling Facilities," EPA/530-SW-88-035, August 1988.
3. Chemical Manufacturers Association, *1986 CMA Hazardous Waste Survey*, prepared by Tischler/Kocurek, Round Rock, Texas, Washington, DC, May 1988.
4. Westat, Inc., *National Survey of Hazardous Waste Generators and Treatment, Storage, and Disposal Facilities Regulated Under RCRA in 1981*, Office of Solid Waste, U.S. Environmental Protection Agency, Washington, DC, April 1984.
5. Congressional Budget Office, *Hazardous Waste Management: Recent Changes and Policy Alternatives*, U.S. Government Printing Office, May 1985.
6. Office of Technology Assessment, *From Pollution to Prevention: A Progress Report on Waste Reduction*, U.S. Government Printing Office, June 1987.
7. Committee of Institutional Considerations in Reducing the Generation of Hazardous Industrial Wastes, *Reducing Hazardous Waste Generation: An Evaluation and a Call for Action*, National Academy Press, Washington, DC, 1985.
8. Ventura County Department of Environmental Health, *Hazardous Waste Reduction Guidelines for Environmental Health Programs*, State of California, Department of Health Services, Toxic Substance Control Division, Alternative Technology Section, Sacramento, California, May 1987.
9. U.S. Environmental Protection Agency, *Waste Minimization: Issues and Options (Volume One)*, Office of Solid Waste, Washington, DC, October 1986.

10. U.S. Environmental Protection Agency, *Waste Minimization: Issues and Options (Volume Two)*, Office of Solid Waste, Washington, DC, October 1986.
11. U.S. Environmental Protection Agency, *Waste Minimization: Issues and Options (Volume Three)*, Office of Solid Waste, Washington, DC, October 1986.
12. Congressional Budget Office, *Empirical Analysis of U.S. Hazardous Waste Generation, Management, and Regulatory Costs*, Staff Working Paper, U.S. Congress, Washington, DC, October 1985.
13. David Sarokin, Warren Muir, Catherine Miller, and Sebastian Sperber, *Cutting Chemical Wastes: What 29 Organic Chemical Plants Are Doing To Reduce Hazardous Wastes*, INFORM, New York, 1985.
14. Battelle Columbus Laboratories, *Report on Hazardous Waste Management Needs Assessment*, prepared for the Tennessee Valley Authority, Columbus, Ohio, June 1984.

Waste Reduction Techniques and Technologies

Gary E. Hunt

Pollution Prevention Program
North Carolina Department of Natural Resources and
Community Development
Raleigh, North Carolina

3.1 Introduction to Waste Reduction Techniques

Waste reduction techniques can be applied to any manufacturing process, from something as simple as making a paper clip to something as complex as assembling the space shuttle. Available techniques range from easy operational changes to state-of-the-art recovery equipment.

The common factor in these techniques is that they reduce bottom-line operational costs.

Waste reduction is not a new concept—it has been around as long as humans have been producing products. While it has been known over the years by a number of names, it is merely optimization of the production process. For example, a late-eighteenth-century electroplating manual states that "nothing whatever should be allowed to go to waste in well-conducted works."[1] The manual includes many waste reduction methods that are applicable and are discussed in the literature even today. This points up the fact that many waste reduction techniques are relatively "low-tech." In fact, many industries find that simple operational changes, increased training, and improved inventory management can significantly reduce waste generation rates.

Waste reduction techniques can be broken down into four major categories, as shown in Table 3.1. Because the classifications are broad there will be some overlap. In actual application, waste reduction techniques generally are used in combination so as to achieve maximum effect at the lowest cost.

The selection of specific reduction techniques by an individual business must be based on accurate and current information on waste-stream generation and waste management costs. This is accomplished by developing and implementing a waste reduction program as a key part of a comprehensive waste management plan. Components of the reduction program include procedures for collecting information, evaluating options, and identifying cost-effective waste reduction techniques. Once identified, the techniques can be implemented and become an established part of the facilities management and operation. Approaches to developing such a program are discussed in Chap. 4.

TABLE 3.1   Categories of Waste Reduction Techniques

Inventory management
  Inventory control
  Material control
Production process modification
  Operation and maintenance procedures
  Material change
  Process equipment modification
Volume reduction
  Source segregation
  Concentration
Recovery
  On-site recovery
  Off-site recovery

One point of caution: An evaluation should be made of a reduction technique's impact on all wastestreams, not just the targeted one. This thorough evaluation must be done before the technique is implemented. For example, while changing from a solvent to a water-based cleaner will eliminate the generation of hazardous waste, it may also increase the wastewater's organic load, possibly to such an extent that the facility cannot meet its discharge limits without incurring significantly increased wastewater treatment costs.

3.2 Inventory Management

Proper control over raw materials, intermediate products, final products, and the associated wastestreams is now being recognized by industry as an important waste reduction technique.[2] In many cases waste is just out-of-date, off-specification, contaminated, or unnecessary raw materials, spill residues, or damaged final products. The cost of disposing of these materials not only includes the actual disposal costs but also the cost of the lost raw materials or product. This can represent a very large economic burden on any company. For example, one furniture company had to spend thousands of dollars, at $5/gal, to properly dispose of two-years' worth of unused coating materials. This expense was in addition to the $7/gal the material originally had cost.[3]

There are two basic aspects to inventory management: controlling the types and quantities of materials in the plant inventory; and controlling the handling of raw materials, along with the finished products and wastestreams in the production facility. The former aspect, referred to as *inventory control*, includes techniques to reduce inventory size and hazardous chemical use while increasing inventory turnover. The latter aspect, referred to as *material controls*, includes methods to reduce raw material and finished product loss and damage during handling, production, and storage.

Any effective inventory management program must include process waste. Handling waste as if it were a product will help reduce waste and increase the potential for recovery. Many of the techniques discussed in this section can be applied to waste material as well as to raw materials and finished products.

3.2.1 Inventory control

Methods for controlling inventory range from a simple change in ordering procedures to the implementation of just-in-time (JIT) manufacturing techniques. Most of these techniques are well known in the business community; however, their use as very effective waste reduction techniques has not been widely recognized. Many companies can help reduce their waste generation by tightening up and expanding current inventory-control programs. This approach will significantly impact the three major sources of waste resulting from improper inventory control: excess, out-of-date, and no-longer-used raw materials. For example, a manufacturer of polyvinyl chloride products reduced the quantity of out-of-date and off-specification raw materials generated by over 50 percent through inventory control. Techniques used included purchase of containerized rather than bulk materials, reduction in purchase quantities, and separation and reuse of excess materials where possible. The program took six months to implement at a negligible cost and saved $50,000 per year in raw material and waste management costs.[4]

Purchasing only the amount of raw materials needed for a production run or a set period of time is one of the keys to proper inventory control. Excess inventory often results from a purchasing department getting a "good deal" on a chemical and buying a tank-car load when only a drum is needed. The excess must be disposed of because it goes out of date before it can be used. Better application of existing inventory management procedures should help to reduce this problem and should be coupled with education programs for purchasing personnel on the problems and costs of disposing of excess materials. Additionally, the set expiration dates on materials should be evaluated, especially for stable compounds, to see if they are too short. A furniture manufacturer reduced excess inventories by having one person assigned to the job of purchasing all solvents and finishes for all divisions. Purchases were based on long-term production schedules, which in turn had been developed to fully utilize finishing materials.[5] Another company, a large paint formulator, reduced the quantity of discontinued finished product by developing a computerized procedure to search inventory at 2 plants and 27 warehouses for available stock before formulating another batch. Transportation costs for moving the finished products are small compared to the high hazardous waste disposal costs.[6]

Another approach to inventory control is to purchase the material in the proper amount and the proper size container. If large quantities of a material are used, then purchasing it in bulk will produce less waste, both in product loss and empty packaging, than if it were purchased in drums or bags. On the other hand, small containers may be better than bulk purchases if the material has a short shelf life or is not used in large amounts. Some companies are purchasing material in returnable, reusable containers, thus eliminating the generation of empty bags or containers. In some cases, raw materials such as pigments and biocides can be packaged in small soluble bags which allow the material and container to be put right into the processed product.[7,8]

If surplus inventories do accumulate, steps should first be taken to use the excess material within the plant or company. If this is not successful, then the supplier should be approached to see if it will take the material back. If the supplier won't, the next step is to identify possible users or markets outside the company. Only if this fails should other management options be examined. The subject of market development is covered more fully in Sec. 3.5.

The ultimate in inventory control procedures is JIT manufacturing, since this system eliminates the existence of any inventory by directly moving raw materials from the receiving dock to the manufacturing area for immediate use. The final product is then shipped out without any intermediate storage. JIT manufacturing is a complex program to implement and cannot be used by all facilities; however, when applicable, it can reduce waste significantly. For example, using JIT techniques, the 3M Company reduced waste generation by 25 to 65 percent in its individual plants.[2]

Developing review procedures for all material purchased is another step in establishing an inventory control program. Standard procedures should require that all materials be approved prior to purchase and checked before acceptance at the facility. In the approval process all production materials should be evaluated to determine if they contain hazardous constituents, and if so, what alternative nonhazardous substitute materials are available (see Sec. 3.3.2). Also, brief evaluation procedures should be established for all incoming raw materials to avoid the acceptance of wrong, off-specification, or defective materials. This will reduce the potential for generation of off-specification products, damaged process equipment, and disposal of unusable raw materials.

Development of review procedures determination can be made either by one person with the necessary chemistry background or a committee made up of people with a variety of backgrounds. Needed information can be obtained from the Material Safety Data Sheets (MSDS) provided by the chemical supplier. If these data sheets do not contain enough information, the supplier should be contacted and asked to provide the necessary information in confidence. If the chemical supplier is unable or unwilling to provide the information, either a new supplier should be found or the materials should be chemically analyzed. Any material which has been approved can be ordered; new material must first go through the approval process. This approach fits very well with health and environmental regulations, which require chemical inventories and collection of MSDSs. Evaluation of this information can lead to fewer chemicals being stored, used, and released, thus reducing regulatory burdens and potential liabilities. Many kinds of companies, hailing from the electronics industry to the textile industry, have established successful material review programs.[9,10,11] Material review procedures should also be applied during new-product development. Before a new product is produced, a thorough evaluation of the materials and processes used to make the product should be made. The use of hazardous materials can then be reduced as much as possible prior to beginning production.[12,13]

3.2.2 Material control

An important area commonly overlooked or not given proper attention in many manufacturing facilities is material control procedures, which includes storage of raw materials, products, and process waste and the transfer of these items within the process and around the facility. Proper material control procedures will ensure that raw materials will reach the production process without loss through spills, leaks, or contamination. It also will ensure that the material is efficiently handled and used in the production process and does not become waste. Examples of potential sources of material loss are shown

TABLE 3.2   Potential Sources of Process Material Loss

Area	Source
Loading	Leaking fill hose or fill line connections
	Draining of fill lines between filling
	Punctured, leaking, or rusting containers
	Leaking valves, piping, and pumps
Storage	Overfilling of tanks
	Improper or malfunctioning overflow alarms
	Punctured, leaking, or rusted containers
	Leaking transfer pumps, valves, and pipes
	Inadequate diking or open drain valve
	Improper material transfer procedures
	Lack of regular inspection
	Lack of training program
Process	Leaking process tanks
	Improperly operated and maintained process equipment
	Leaking valves, pipes, and pumps
	Overflow of process tanks; improper overflow controls
	Leaks and spills during material transfer
	Inadequate diking
	Open drains
	Equipment and tank cleaning
	Off-specification raw materials
	Off-specification products

in Table 3.2. A resin manufacturer provides an example of how proper raw-material transfer procedures can reduce waste generation. With this manufacturer it was common practice to let the residual material in the hose from the phenol delivery trucks drain into the plant's wastewater collection system after the storage tanks were filled. The unloading procedures were then changed to require the hoses to be flushed with water and the phenol-water mixture be stored in a tank for use in the production process.[14]

Material loss can be greatly reduced through improved process operation, increased maintenance, and additional employee training. Many sources of material loss, such as leaks and spills, can be easily identified and corrected. For example, a dairy plant reduced its product loss by 65 percent through an improved maintenance program.[15] Many of the techniques which can be used to reduce material loss are discussed in Sec. 3.3.1.

Another problem area which is frequently overlooked is waste handling procedures. Waste should be handled and managed like a product. Allowing a recyclable or clean waste material to be contaminated may reduce or eliminate any recovery potential. One way to improve the material control procedures is to have at least one person responsible for handling and tracking waste materials within the plant. This approach can be very cost-effective when the waste material is recoverable or has some market value. This area is discussed further in Sec. 3.5.

3.3 Production Process Modification

Improving the efficiency of a production process can significantly reduce waste generation. Use of this approach can help reduce waste at the source of generation, thus decreasing waste management liability and costs. Some of the most cost-effective reduction techniques are included in this category; many are simple and relatively inexpensive changes to production procedures. Available techniques range from eliminating leaks from process equipment to installing state-of-the-art production equipment. The waste reduction techniques in this category can be divided into improved operation and maintenance, material change, and equipment modifications. Each topic is discussed in more detail in the following sections.

3.3.1 Operational and maintenance procedures

Significant amounts of waste can be reduced through improvements in the way a production process is operated and maintained. This approach is one of the most overlooked of all waste reduction areas because many wasteful operational practices have gone on so long they have become standard operating procedures. Also, in many cases, maintenance is so busy correcting current problems that preventive maintenance is overlooked until it is too late. Improvements in operation and maintenance usually are relatively simple and cost-effective. Most of the techniques are not new or unknown. However, a good deal of information on the source and cause of the waste generation must be known before effective measures can be developed.

A certain dairy processing facility provides a good example of significant waste reduction through improved operation and maintenance procedures. A milk loss prevention program was instituted which included employee training on proper operations, causes of product loss, and impact of waste generation; increased maintenance, including correction of current leaking equipment; better tracking of product losses; and development of improved operational techniques. This program saved the company $288,000 per year and has resulted in a significant reduction in the organic content of the wastewater. All this was obtained without any capital costs.[15,16]

Operational procedures   A wide range of methods are available to operate a production process at peak efficiency. These methods are neither new nor unknown and are usually inexpensive to institute, as little or no capital cost is necessary. For example, a manufacturer of plumbing fixtures changed the concentration of chrome in its electroplating baths to the low end of the recommended operating range. By reducing the chrome concentration from 3700 mg/liter to 3350 mg/liter the amount of chrome which had to be treated was reduced by 9 percent without affecting product quality. This not only saved raw material and treatment chemicals, but it also reduced wastewater treatment sludge generation.[17]

Improved operation procedures are quite simply methods which make optimum use of the raw materials used in the production process. Some examples are shown in Table 3.3. Most production processes, no matter how long they have been in operation or how well they are run, can be operated more efficiently. Many sources of waste generation become overlooked because "that is just the way the process works." Additionally, some process steps may in fact be unnecessary, and eliminating them will reduce waste generation. For example, a paint manufacturer found that instead of using a series of coarse to fine filters for grit removal, only the fine filter was necessary. As a result of this change to using only one filter, the generation of spent filter cartridges was reduced by 50 percent.[6] Another company discovered that standard operating procedure was to degrease all parts coming into the facility and then re-oil those that did not require further

TABLE 3.3 Examples of Operational Changes to Reduce Waste Generation

Reduce raw material and product loss due to leaks, spills, drag-out, and off-specification process solution.

Schedule production to reduce equipment cleaning. For example, formulate light to dark paint so the vats do not have to be cleaned out between batches.

Inspect parts before they are processed to reduce number of rejects.

Consolidate types of equipment or chemicals to reduce quantity and variety of waste.

Improve cleaning procedures to reduce generation of dilute mixed waste with methods such as using dry cleanup techniques, using mechanical wall wipers or squeegees, and using "pigs" or compressed gas to clean pipes and increasing drain time.

Segregate wastes to increase recoverability.

Optimize operational parameters (such as temperature, pressure, reaction time, concentration, and chemicals) to reduce by-product or waste generation.

Develop employee training procedures on waste reduction.

Evaluate the need for each operational step and eliminate steps that are unnecessary.

Collect spilled or leaked material for reuse.

SOURCE: From Refs. 4, 6, 7, 19, and 20.

coating and that cleaning only those parts which needed further processing would reduce waste solvent generation.[18] Thus, asking why a process step has to be done will help identify reduction possibilities.

These approaches can be used by all sizes and types of facilities, even by well-designed and well-operated complex operations. For example, an organic chemical company, during a comprehensive process evaluation, found three significant sources of isobutylene emissions to the atmosphere. Laboratory scale tests found that changes in the reaction conditions (temperature, residence time, and concentration) would almost eliminate isobutylene emission while also increasing product generation. For example, for one reaction step emissions were reduced 99 percent, primary product yield was increased from 94.5 to 99.7 percent, and batch cycle time was reduced by half. The total saving from the operational changes at all these reaction steps was $500,000 per year, with a total capital cost of $80,000. Additionally, the need for air pollution control equipment was eliminated.[20]

Once proper operating procedures have been established they must be fully documented and made part of the employee training program. A comprehensive training program is a key element of any effective waste reduction program. For example, a dairy reduced waste by 14 percent and a semiconductor manufacturer reduced waste by 40 percent through use of such training programs.[18,22] For a program to be effective, all levels of personnel should be included, from the line operator to the corporate executive officer. The goal of any program is to make the employee aware of waste generation, its impact on the company and the environment, and ways it can be reduced. Top level managers have to be made aware of the costs, problems, and liabilities of waste management, and the economic benefits of waste reduction. Line managers must understand the effect of their process line on waste generation in order to focus on how waste can be reduced, how to educate and motivate their employees. The majority of the training program should be directed at the line operators. Some possible program elements are shown in Table 3.4. The training program should emphasize management's commitment to waste reduction, the beneficial impact of reduction on job security, and the improvements reduction will make to the local environment. All employees should attend refresher sessions on a regular basis. All new workers should undergo the training, which could be done as part of an established worker right-to-know training program for new employees.

Companies have taken a number of different approaches to waste reduction training. One facility has developed a training video to introduce waste reduction to the line operator with simple written waste reduction procedures for each line operation. Also at this facility, training and written material were developed for line supervisors and all levels of management.[18] A furniture manufacturer took a rather unique approach. Working closely with the line operator, video equipment was used to record the operator's spray technique. The tape was later reviewed by the operator and training personnel to identify problem areas and provide corrective procedures. The operator was videotaped again later so that the operator could see the improvements. The company estimates that coating use and the associated waste gen-

---

TABLE 3.4   Elements in a Waste Reduction Training Program

1. Explain the need for waste reduction and emphasize benefits to employee and community.

2. Explain the direct effect that an employee can have on improved work and living environment.

3. Express management's commitment to waste reduction.

4. Explain waste management terminology in simple terms.

5. Present general overview of environmental regulations which impact the facility.

6. Examine improved operational practices for reducing waste generation. Illustrate good and poor operating practices utilizing slides or video. Use positive language, e.g., "this is a better way of doing things," instead of "this is what you have to do."

7. Solicit ideas for waste reduction methods and explore possible solutions to identified problems.

SOURCE: From Ref. 23.

eration costs were reduced by about 10 percent. This represented a savings of about $60,000 per year in coating material costs alone. Savings were also realized through reduced waste from spray booth cleanout and reduced air emissions.[3]

Maintenance programs One company stated that one-fourth to one-half of its excess waste load was due to poor maintenance.[24] A strict maintenance program which stresses corrective *and* preventive maintenance can reduce waste generation caused by equipment failure. Such a program can help spot potential sources of release and correct a problem before any material is lost. A good maintenance program is so important because the benefits resulting from the best waste reduction program can be wiped out by just one process leak or equipment malfunction.

A maintenance program can include maintenance cost tracking and preventive maintenance scheduling and monitoring. To be effective, a maintenance program should be developed and followed for each operational step in a production process, with special attention to potential problem points. A strict schedule and accurate records on all maintenance activities should be maintained. The type of information which should be collected and updated regularly in order to establish a preventive maintenance program is listed below:[25]

- A list of all plant equipment and location
- Operating time for each item or area
- Which items are critical to the process(es)
- Problem equipment
- Previous maintenance history
- Vendor maintenance manuals
- A data base of equipment repair histories

Computer-based maintenance scheduling and tracking programs are available from a variety of vendors. A comprehensive program can also include predictive maintenance. This approach provides the means to schedule repairs or replacement of equipment based on actual condition of the machinery. A number of nondestructive testing technologies are available for making the necessary evaluations for this approach.[26]

Maintenance procedures themselves produce waste, such as process materials, rags, scrap parts, oils, and cleanup residue. These wastes, too, can be reduced by using standard waste reduction techniques such as revised operational procedures, equipment modification, source segregation, and recovery. For example, before a filter is replaced, all process material should be drained from the housing, either under gravity or pressure, and collected for reuse.[27]

3.3.2 Material change

Hazardous material used in either a product formulation or in a production process may be replaced with a less hazardous or nonhazardous material. Reformulating a product to contain less hazardous material should reduce the amount of hazardous waste generated during both the product's formulation and its end use. Using a less hazardous material in a production process will generally reduce the amount of hazardous waste produced. Some examples of material change to reduce waste generation are given in Table 3.5.

Product reformulation is one of the more difficult waste reduction techniques, yet it can be very effective. As more manufacturers implement inventory management programs, pressure will increase on chemical supply companies to produce products with lower quantities of hazardous materials. Due to the proprietary nature of product formulations, specific examples of product reformulation are scarce. General examples include eliminating pigments containing heavy metals from ink, dyes, and paint formulations; replacing chlorinated solvents

TABLE 3.5 Examples of Waste Reduction Through Material Change

Industry	Technique
Household appliances	Eliminate cleaning step by selecting lubricant compatible with next process step.[28]*
Printing	Substitute water-based ink for solvent-based ink.[29]
Textiles	Reduce phosphorus in wastewater by reducing use of phosphate-containing chemicals.[30] Use ultraviolet light instead of biocides in cooling towers.[35]
Air conditioners	Replace solvent-containing adhesives with water-based products.[31]
Electronic components	Replace water-based film-developing system with a dry system.[9]
Aerospace	Replace cyanide cadmium-plating bath with a noncyanide bath.[32]
Ink manufacture	Remove cadmium from product.[33]
Plumbing fixtures	Replace hexavalent chrome-plating bath with a low-concentration trivalent chrome-plating bath.[34]
Pharmaceuticals	Replace solvent-based tablet-coating process with a water-based process.[36]

*Superior numbers refer to sources in Reference list at end of chapter.

with nonchlorinated solvents or water in cleaning products; replacing phenolic biocides with other less toxic compounds in metalworking fluids; and developing new paint, ink, and adhesive formulations based on water rather than organic solvents. The applicability of this technique will be very product-specific, but with the ever-increasing cost and liability associated with waste management, it is a very important waste reduction and business strategy.

Hazardous chemicals used in the production process can also be replaced with less hazardous or nonhazardous materials. Changes can range from using purer raw materials to replacing solvents with water-based products, which is a very widely used waste reduction technique that is applicable to many industries. Many of these changes involved switching from a solvent to a water-based process solution. For example a diesel engine remanufacturing facility switched from cleaning solvents and oil-based metalworking fluids to water-based products. This change reduced their coolant and cleaning costs by about 40 percent. Additionally, the company was able to eliminate a cleaning step, and machine filters lasted twice as long, thus reducing material and labor costs.[37] A material change may also allow the elimination of a process step. A manufacturer of home appliances used this approach to replace an alkaline degreasing step used in a stamping process. A water-based degreaser had replaced a solvent degreaser using 1,1,1-trichloroethane. However, the alkaline degreaser residue on the parts caused deterioration of the stamping dies. The cleaning step was then eliminated by selecting a new lubricant which was noncorrosive to the stamping dies. Additionally, this lubricant did not have to be removed before the next process step, annealing, because it would be burned off in the annealing ovens.[28]

This last example points out one major source of possible problems with material changes, i.e., an adverse effect on the production process, product quality, or waste generation. This possibility can be overcome by carefully evaluating possible impacts of the proposed change on worker health, product quality, process operation, inventory, production costs, and waste management. Conducting pilot-scale tests, trial runs, and phasing in the change over time will help eliminate some potential problems.

One important area which is sometimes overlooked in making a material change is its impact on the total wastestream. Switching from a solvent-based to a water-based product can increase wastewater volumes and concentrations, which could adversely affect the current wastewater treatment system, cause effluent limits to be exceeded, and possibly increase wastewater treatment sludge production. Thus, before any change is made, its impact on all emissions must be evaluated. For example, a small metal fabricator switching from a solvent-based to a water-based product can significantly increase the organic and oil concentration in the wastewater. This increase may require the plant to install new or additional wastewater treatment capacity and/or pay significantly increased sewer use fees.

Reducing or eliminating hazardous materials from the production process can decrease not only hazardous waste generation, but also the quantity of hazardous materials in air emissions and wastewater effluents. This can, if properly evaluated, reduce the capital investment in treatment systems needed to meet pollution discharge limits. For example, a producer of gift-wrapping paper switched from solvent-based to water-based inks. This change saved the company $35,000 per year on hazardous waste management costs. Furthermore, it saved the company from having to install an air pollution control system, to control volatile organic carbon emissions, which would have cost several million dollars.[31]

3.3.3 Process equipment modification

Waste generation may be reduced by installing more efficient process equipment or modifying existing equipment to take advantage of better production techniques. New or updated equipment can use process materials more efficiently, producing less waste. Additionally, higher-efficiency systems may reduce the number of rejected or off-specification products, thereby reducing the amount of material which has to be reworked or disposed of. The use of more efficient equipment or processes can pay for itself through higher productivity, reduced raw material costs, and reduced waste management costs. The necessary capital investment can usually be justified by the increased production rates, and reduced waste management costs provide an added bonus.[33,9] Some examples of process modifications are given in Table 3.6.

Modifying existing process equipment can be a very cost-effective method for reducing waste generation. In many cases the modifications can just be relatively simple and inexpensive changes in the way the materials are handled within the process to ensure that they are not wasted or lost. This can be as easy as redesigning parts racks to reduce drag-out in electroplating operations, installing better seals on process equipment to eliminate leakage, or installing drip pans under equipment to collect leaking process material for reuse.[38,18] In many cases, process modifications and improved operational procedures are used together to reduce waste. One chemical company reduced the waste from a sump in a production area from 31,750 kg/year to 1360 kg/year by installing a sight glass, using better pump scales, and purchasing a broom.[39] Modifying equipment to improve operational effi-

TABLE 3.6   Examples of Production Process Modifications for Waste Reduction

Process step	Technique
Chemical reaction	Optimize reaction variables and improve process controls.
	Optimize reactant-addition method.
	Eliminate use of toxic catalysts.
	Improve reactor design.
Filtration and washing	Eliminate or reduce use of filter aids and disposable filters.
	Drain filter before opening.
	Use countercurrent washing.
	Recycle spent washwater.
	Maximize sludge dewatering.
Parts cleaning	Enclose all solvent cleaning units.
	Use refrigerated freeboard on vapor degreaser units.
	Improve parts draining before and after cleaning.
	Use mechanical cleaning devices.
	Use plastic-bead blasting.
Surface finishing	Prolong process bath life by removing contaminants.
	Redesign part racks to reduce drag-out.
	Reuse rinse water.
	Install spray or fog nozzle rinse systems.
	Properly design and operate all rinse tanks.
	Install drag-out recovery tanks.
	Install rinse water flow control valves.
	Install drip racks and drainboards.
Surface coating	Use airless air-assisted spray guns.
	Use electrostatic spray-coating system.
	Control coating viscosity with heat units.
	Use high-solids coatings.
	Use powder coating systems.
Equipment cleaning	Use high-pressure rinse system.
	Use mechanical wipers.
	Use countercurrent rinse sequence.
	Reuse spent rinse water.
	Use "pigs" to clean lines.
	Use compressed gas to blow out lines.
Spills and leaks	Use bellow-sealed valves.
	Install spill basins or dikes.
	Use sealless pumps.
	Maximize use of welded pipe joints.
	Install splash guards and drip boards.
	Install overflow control devices.

SOURCE: From Refs. 18, 21, and 40–44.

ciency requires a thorough understanding of both the production process and wastestream generation. In many cases it is best to phase in the required process modifications over a period of time to reduce any potential effects on the whole production process and to evaluate its impact on the wastestream. As discussed before, the impact on all wastestreams must be evaluated.

Installing new, more efficient equipment and, in some cases, modifying current equipment will require capital investment in equipment, facility modifications, and employee training. The extent of the investment will vary over a large range depending on the type of equipment used. Replacing a solvent vapor degreaser with a household dishwasher to clean circuit boards may cost only several hundred dollars, while replacing a spray coating operation with a powder coating system may cost hundreds of thousands of dollars and require new facilities and extensive employee training. What they have in common is that they both will pay back the investment costs.[18]

Examples of new, more efficient process equipment are numerous in the literature, but little is usually said about reduced waste generation and management costs. The following example shows what one company saved when new process equipment was installed. When an automated metal electroplating system was installed to replace the manual operation, annual productivity increased and system downtime decreased from 8 percent to 4 percent. Chemical consumption decreased 25 percent, resulting in an annual reduction of $8000 per year in raw material costs. Water costs were reduced by $1100 per year, and plating wastes, including acids, caustic, and oils, decreased from 204 kg/day to 163 kg/day. Treatment costs for the process water used in the plating operation were reduced by 25 percent. Annual personnel and maintenance cost savings attributable to the new system are $35,000 per year. The automated system has also eliminated worker exposure to acids and caustics, which had been required with the previous manual operation.[17]

As discussed earlier in this section, the use of more efficient process equipment can also mean changing to less hazardous process materials. For example, one furniture company switched from a manual staining operation using solvent-based materials to an automatic staining system using water-based stains. The new process reduced the time required to stain a pallet of parts by 95 percent and reduced raw material usage by 20 percent. The savings in labor alone paid for the new staining system in just three months.[3]

One important factor which is sometimes overlooked in evaluating the cost-effectiveness of equipment modifications is the cost associated with reworking or disposing of off-specification products. This can be very expensive, not only in terms of labor and materials, but also in waste management costs. In many manufacturing operations which involve coating a product, such as electroplating or painting, chemicals are used to strip off the coating from rejected products so that they can be recoated. These chemicals, which can include acids, caustics, cyanides, and chlorinated solvents, often are hazardous wastes which must be properly managed. By reducing the quantity of parts which must be reworked, the quantity of waste can be reduced.

3.4 Volume Reduction

Volume reduction includes techniques to separate hazardous wastes and recoverable wastes from the total wastestream. These techniques are usually used to increase recoverability, reduce the volume and thus the disposal costs, or increase management options. The available techniques used range from simple segregation of wastes at the source to complex concentration technology, as shown in Table 3.7. They are applicable to all types of wastestreams. These techniques can be divided into two general areas: source segregation and waste concentration. Only those methods which are actual waste reduction techniques will be discussed in this section.

3.4.1 Source segregation

Segregation of wastes is in many cases a simple and economical technique for waste reduction. For example, by segregating wastes at the

TABLE 3.7   Examples of Waste Reduction Through Volume Reduction

Industry	Technique
X-ray film	Segregate polyester film scrap from other production waste and recycle.[2]*
Resins	Collect waste resin and reuse in next batch.[62]
Printed circuit boards	Use filter press to dewater sludge to 60% solids and sell sludge for metal recovery.[17]
Pesticide formulation	Use separate bag houses at each process line and recycle collected dust into product.[45]
Research laboratory	Segregate chlorinated and nonchlorinated solvents to allow off-site recovery.[17]
Aircraft components	Use ultrafiltration to remove recoverable oil from spent coolants.[13]
Paint formulation	Segregate and reuse tank-cleaning solvents in paint formulations.[23]
Furniture	Segregate and reuse solvents used to flush spray-coating lines and pumps as coating thinner.[33]

*Superior numbers refer to sources in Reference list at end of chapter.

source of generation and handling the hazardous and nonhazardous wastes separately, waste volume and thus management costs can be reduced. Additionally, the uncontaminated or undiluted wastes may be reusable in the production process or may be sent off site for recovery.

The segregation technique is applicable to a wide variety of wastestreams and industries and usually involves simple changes in operational procedures. For example, in metal-finishing facilities, wastes containing different types of metals can be treated separately so that the metal values in the sludge can be recovered. Keeping spent solvents or waste oils segregated from other solid or liquid waste may allow them to be recycled. Wastewater containing toxic material should be kept separate from uncontaminated process water to reduce the overall volume of water which must be treated.

A commonly used waste segregation technique is to collect and store washwater or solvents used to clean process equipment (such as tanks, pipes, pumps, or printing presses) for reuse in the production process. This technique is used by paint, ink, and chemical formulators, as well as by printers and metal fabricators. For example, a printing firm segregates and collects toluene used for press and roller cleanup operations. By segregating the used toluene by color and type of ink contaminant, it can be reused later for thinning the same type and color of ink. The firm now recovers 100 percent of the waste toluene, totally eliminating a hazardous wastestream.[17]

Another way to apply the segregation technique is to collect and reuse back in the product dust and excess materials generated during the manufacturing process. This technique is used by one pesticide formulator to reduce the volume of hazardous waste it generates. The firm collects the dust generated during the formulation process in a separate bag house for each process line. The collected dust is then reused in the product as an inert filler. This has eliminated the generation of 20,412 kg/year of waste, saving the company $9000 in disposal costs and $2000 in raw material costs. These savings paid for the necessary equipment in less than a year.[17]

3.4.2 Concentration

Various techniques are available to reduce the volume of a waste through physical treatment. Such techniques usually remove a portion of a waste, such as water. For example, concentration techniques are commonly used to dewater wastewater treatment sludges and reduce the volume by as much as 90 percent. Available concentration methods include gravity and vacuum filtration, evaporation, ultrafiltration, reverse osmosis, freeze vaporization, filter press, heat drying, and compaction. Many of these actually are recovery techniques and will be discussed further in the next section. Concentration techniques are available for all types of wastestreams and are used by a wide range of industries.

Unless a material can be recycled, just concentrating a waste so more waste can fit into a drum is not waste reduction. In some cases concentration of a wastestream may also increase the likelihood that the material can be reused or recycled. For example, filter presses or sludge driers can increase the concentration of metals in electroplating wastewater treatment sludge to such a level that they become valuable raw material for metal smelters. A printed circuit board manufacturer dewaters its sludge using a filter press to 60 percent solids. The company receives $7200 per year through the sale of the dewatered sludge for copper reclamation.[17]

3.5 Recovery

Recovering wastes can provide a very cost-effective waste management alternative. This technique can help eliminate waste disposal costs, reduce raw material costs, and possibly provide income from a salable waste. Recovery of wastes is a widely used practice in many manufacturing processes and can be done on site or at an off-site facility. It has been used for centuries to increase productivity and profitability. In fact, the organic chemical industry in the mid-1800s produced aniline dyes which were made from coal gasification plant waste.[46] Current production facilities have the potential to increase greatly the recycling of waste materials based on present regulatory impacts, increased waste management costs, and new waste recovery technologies and approaches.

Waste recovery should only be considered after all other waste reduction options have been instituted. Actually reducing the amount of waste generated at the source will usually be more cost-effective than recycling. In many cases this is because a waste is lost raw material or product which requires time and money to manage and recover. Also, simply generating and handling a waste produces a range of regulatory, health, and environmental liabilities.

All of the previously discussed waste reduction techniques can be used in conjunction with recovery to produce a cost-effective waste management program. One company, a producer of breaded foods, used a variety of waste reduction techniques to significantly reduce waste generation. A management and employee training program on waste reduction techniques and their impact on company profits was instituted. Improved operation and maintenance procedures were instituted, including using dry cleanup for spills and equipment, installing or modifying drip trays under process equipment to better collect lost process material, and developing better systems for collecting and handling waste materials so they could be sold to a recovery firm. These, plus a number of reduction techniques, decreased water usage by about 30 percent, eliminated the landfilling of waste solids, reduced the organic load of the wastewater by almost 80 percent, and allowed the company to sell 2,359,000 kg/year of solids to recovery firms.[47]

The effective use of recovery will depend on the segregation of the recoverable waste from other process wastes or extraneous material. This segregation ensures that the waste is uncontaminated and the concentration of recoverable material is maximized. The waste, then, must be handled like a product. Many of the inventory control techniques discussed in Sec. 3.1 can be applied to waste materials. Some companies have assigned responsibility for the handling, collection, and scheduling of recovery of waste material to one individual.[13,47,48] This helps ensure that the maximum value of the waste can be recovered.

3.5.1   On-site recovery

In most cases the best place to recover process wastes is within the production facility. Wastes which are simply contaminated versions of the process raw materials are good candidates for in-plant recycling. Such recovery can significantly reduce raw material purchases and waste disposal costs. Waste can be most efficiently recovered at the point of generation, because the possibility of contamination with other waste materials is reduced, as is the risk involved with handling and transporting waste materials. See Sec. 3.4 for information on segregation techniques. Some examples of on-site waste recovery are shown in Table 3.8.

Some wastestreams can be reused directly in the original production process as raw material. This easy reuse is usually accomplished when the waste material is lightly contaminated or is excess raw material. Examples include the cleaning waste from printers, coaters, and chemical or product formulators; electroplating drag-out solutions; process solutions from filter changes; and dust collector residue from pesticide formulators.[21,27,38,45] Lightly contaminated waste can sometimes be reused in operations not requiring high-purity materials. For example, spent high-purity solvents generated during the production of microelectronics can be reused in less critical metal-degreasing operations, or a caustic waste material can be reused to treat an acid wastestream.

Some waste may have to undergo some type of purification before it

TABLE 3.8   Examples of Waste Reduction Through Recovery and Reuse

Industry	Technique
Printing	Use a vapor-recovery system to recover solvents.[17]*
Photographic processing	Recover silver, fixer, and bleach solutions.[17]
Metal fabrication	Recover synthetic cutting fluid using a centrifuge system.[33]
Mirror manufacturing	Recover spent xylene using a batch-distillation system.[33]
Printed circuit boards	Use electrolytic recovery system to recover copper and tin/lead from process wastewater.[49]
Tape measures	Recover nickel-plating solution using an ion-exchange unit.[40]
Medical instruments	Use reverse-osmosis system to recover nickel-plating solution.[40]
Power tools	Recover alkaline degreasing baths using an ultrafiltration system.[17]
Textiles	Use ultrafiltration system to recover dye stuffs from wastewater.[50]
Hosiery	Reconstitute and reuse spent dye baths.[50]
Food processing	Send all solids off site for by-product recovery.[47]
Wastewater treatment	Reuse waste caustic solids to treat acid wastestream.[51]
Pickles	Transfer waste brine pickle solution to a textile plant as a replacement for virgin acetic acid.[52]
Chemicals	Use spent electrolyte from one division as raw material in another.[53] Purify hydrochloric acid in wastestream and sell as a product.[54]
Industrial and consumer products	Segregate and sell office paper, corrugated cardboard, paper trimming, and rejected paper products.[2]
Aluminum die-caster	Sell waste fumed amorphous silica for use in concrete.[55]

*Superior numbers refer to sources in Reference list at end of chapter.

can be reused. A number of physical and chemical techniques available on the market can be used to reclaim a waste material, as shown in Table 3.8. These techniques range from simple filtration to state-of-the-art techniques such as freeze crystallization.[56,57] The method of choice will depend on the physical and chemical characteristics of the wastestream recovery economics, as well as on operational requirements. For a number of wastestreams there are economical small modular recovery units available. Some of the materials commonly recycled by these modular units are spent solvents, electroplating process waste, and spent metalworking fluids.

A certain manufacturer of shopping carts provides a good example of on-site recovery methods and cost savings. The company installed three atmospheric evaporators on the nickel- and chrome-plating lines. These units evaporate water from the plating bath, allowing all rinse water to be returned and thus recovering all dragged-out nickel and chrome plating solutions. With the start-up of the evaporators, raw chemical use declined dramatically: purchases of 1800 kg of nickel sulfate per month have declined to zero; chromic acid purchases have declined by 75 percent, from 900 kg/month to 200 kg/month; and sodium bisulfite used in the waste treatment process for chromic acid has declined from 1800 kg/month to 200 kg/month. Annual savings are projected to be $112,500. The cost of the three atmospheric evaporators was $60,500, with a payback period of only seven months and an annual savings of $145,000.[9]

Most on-site recovery systems will generate some type of residue (i.e., contaminants removed from the recovered material). This residue can either be processed for further recovery or properly disposed of. For example, a solvent distillation system will usually recover only 90 percent of the input spent solvent for reuse. The residues or distillation bottoms are a hazardous waste and as such must be managed in an approved manner. The economic evaluations of any recovery technique must include the management of these residues.

3.5.2 Off-site recovery

Wastes may be recovered at an off-site facility when (1) the equipment is not available to recover on site, (2) not enough waste is generated to make an in-plant system cost-effective, or (3) the recovered material cannot be reused in the production process. Off-site recovery usually entails recovering a valuable portion of the waste through chemical or physical processes or directly using the waste as a substitute for virgin material. Some materials which are commonly reprocessed off site are oils, solvents, electroplating sludges and process baths, lead-acid batteries, scrap metal, food-processing waste, plastic scrap, and cardboard. Wastes directly used are usually chemically or physically specific for a select purpose (i.e., have the same physical or chemical properties as the raw materials they replace) and can range from concentrated acids to chemical by-product streams. This alternative is described further in Sec. 3.2.1. Some examples of materials which have been recovered off site are shown on Table 3.8.

For wastes regulated under the Resource Conservation and Recovery Act (RCRA) there are regulatory barriers to some forms of off-site recovery. Additionally, for hazardous waste the generator is liable for the waste material handled at a recovery facility and for any residue produced. This responsibility may pose possible long-term liability problems. Check with state and federal regulatory agencies for the impact of the regulation on any off-site recovery activity.

The cost of off-site recycling will depend on the purity of the waste and the market for the waste or recovered material. Some materials may be salable, while others may require a fee to be paid for disposal. The markets for some wastes, such as scrap metals and waste oils, are very volatile, and a waste material which has a value one day may have none the next.

For some process materials, such as solvents, the supplier may provide services to pick up the material, recover it, and return it for reuse. Known as *tolling*, this vendor service provides a good way to reduce disposal and raw material costs because the reclaimed material is usually cheaper than virgin material. This arrangement is widely used for metal-cleaning operations, such as machine shops or automotive repair stations, that use cold cleaners. In these operations, a solvent cleaner is provided for use in a leased parts-washing sink. On a regular schedule the supplier brings in fresh solvent and removes the spent solvent for recovery. This procedure is much cheaper for a small company than if it were to dispose of the solvent itself; however, the generator is still liable for proper recovery of the solvent and management of the distillation residues.

In some situations a waste may be transferred to another company for use as a raw material in the other company's manufacturing process. This exchange can be economically advantageous to both firms as it will reduce the waste disposal costs of the generator and reduce the raw material costs of the user. This upgrade of a waste into a product requires a strong commitment from the generator to find markets for the waste material. In some cases the production process or the waste may have to undergo some modification in order to make a more salable product.

A product-development approach can be used to identify and develop markets for a waste material. An example of the ordered steps in such an approach follows:[53,58]

- Determine waste composition and generation rate.
- Evaluate on-site and in-company uses.
- Evaluate off-site uses and locate potential customers and markets.
- Identify any processing required to meet product specifications.
- Evaluate potential customers.

- Negotiate recycle agreement.

The marketing effort is usually undertaken by individuals in a company's sales or purchasing department, with support from engineering and technical staff. Some larger companies have dedicated groups for just marketing waste materials and surplus materials. Markets for the waste can be identified through advertising, contacts with trade associations, waste exchanges, industrial contacts, brokers, and a number of literature sources.[58,59]

Regional waste exchanges have been set up by a number of states to act as information clearinghouses of wastes that are available and wastes that are sought. A waste exchange can help identify possible markets for a material. The service usually offered is a listing in a catalog or computer data base form of wastes available from generators and sought by users. This information is distributed throughout a specific geographic area. A company interested in a waste contacts the waste exchange, which forwards the inquiry to the listing company. Usually, actual negotiations and material transfers are handled directly by the companies. A current listing of waste exchanges is given in Table 3.9.

Before a company looks outside itself for markets, internal uses should be evaluated. Probably some of the most cost-effective recycling efforts are made within a company because the waste replaces material that otherwise would have to be purchased or produced. It may be possible to use the material directly, or the material may have to undergo some further processing, as discussed below. One chemical manufacturer had the distillation bottoms from a production process refined at an outside custom processor for use as raw material at another facility, saving the company over \$1 million per year in raw material and disposal costs.[53] One approach for excess off-specification material is to set up a corporate waste exchange using a newsletter or even a computer data base system to help identify users for waste material. This approach has been taken by a number of companies.[12,27,59] For example, one large corporation purchasing department lists surplus materials at all plant locations in a computer data system. Before any materials are purchased for a given plant, the purchasing department checks to determine if they are available at another location.[27]

Some wastes may have to be upgraded before a market can be found for them. This upgrading can be in the form of purification, concentration, particle sizing, or other processes. The material itself can undergo processing, or else the production process can be modified to produce a higher-quality wastestream. For example, a circuit board manufacturer found that in order to market its copper-containing wastewater treatment sludge, the sludge must have a low iron concen-

TABLE 3.9   North American Waste Exchanges

Alberta Waste Materials Exchange 4th Floor Terrace Plaza 4445 Calgary Trail South Edmonton, Alberta Canada T6H 5R7	Industrial Waste Information Exchange New Jersey Chamber of Commerce 5 Commerce Street Newark, NJ 07102
California Waste Exchange Department of Health Services Toxic Substances Control Division 714 P Street Sacramento, CA 95814	Manitoba Waste Exchange c/o Biomass Energy Institute 1329 Niakwa Road Winnipeg, Manitoba Canada R2J 3T4
Canadian Waste Materials Exchange Ontario Research Foundation Sheridan Park Research Community Mississauga, Ontario Canada L5K 1B3	Montana Industrial Waste Exchange Montana Chamber of Commerce P.O. Box 1730 Helena, MT 59624
Great Lakes Regional Waste Exchange 470 Market St., S.W., Suite 100A Grand Rapids, MI 49503	Northeast Industrial Waste Exchange 90 Presidential Plaza, Suite 122 Syracuse, NY 13202
Indiana Waste Exchange Environmental Quality Control 1220 Waterway Boulevard P.O. Box 1220 Indianapolis, IN 46206	Southeast Waste Exchange Urban Institute UNCC Station Charlotte, NC 28223
Industrial Material Exchange Service P.O. Box 19276 Springfield, IL 62794-9276	Southern Waste Information Exchange Institute of Science and Public Affairs Florida State University P.O. Box 6487 Tallahassee, FL 32313

SOURCE: From Ref. 60.

tration and be high in solids and copper concentration. An effort was undertaken to modify the wastewater treatment process so that iron-containing treatment chemicals were not used. Additionally, the sludge was dewatered using a filter press to 50 percent solids. This dewatering increased the copper concentration to 25 to 35 percent by weight, and a market for the material with a copper smelter was found. This effort, along with finding uses for several other waste-streams, now allows the company to market over 90 percent of the hazardous waste it generates.[61]

3.6   Summary

As has been shown, a wide range of waste reduction techniques currently exist and are available for most manufacturing steps. However, technology alone will not reduce waste generation. Only a comprehensive waste reduction program will be successful. Such a program should include management commitment, data collection, cost-effective technology selection and implementation, employee training and involvement, and program monitoring. The foundation of any successful program is the evaluation of what wastes are generated and why they are produced. Based on this information, a range of reduction techniques can then be identified and evaluated, and the cost-effective ones implemented.

Sources of specific information on waste reduction techniques are available from a number of places. The best source, however, is discussion with the process operators, who in most cases can identify operations and equipment problems which generate waste. Additional information may be obtained from trade associations and trade journals, as well as government research reports, regulatory agencies, and technical assistance groups.

Industrial trade associations can provide the most detailed and current technical information. Many associations have staff experts with extensive knowledge and experience in waste management. A list of trade associations can be found in Ref. 62. Trade publications are another very good source of information. Many journals contain articles on case studies, current research, vendor information, and suggestions from industrial experts.

Other sources of technical information are federal and state regulatory agencies and technical assistance programs. The U.S. Environmental Protection Agency (U.S. EPA) has established an Office of Pollution Prevention to help promote waste reduction efforts by the regulated community. These efforts include research, education, and technical information. A number of states have established waste reduction technical assistance programs. The level of assistance offered by these programs varies, and the programs may include on-site technical assistance, access to information data base and documents, workshops, referral services, research, and matching grant programs. Current programs are discussed in Chap. 12. Check with your respective state environmental agencies to find out if such a program is available in your locality.

In the final analysis, waste reduction depends on looking at waste in a different way, not as something that inevitably must be treated and disposed of, but for what it really is—a loss of valuable process materials, the reduction of which can have significant economic benefits. One corporate executive summarized it all when he stated that waste is a specialty product for which a market has not yet been found.[63]

3.7 REFERENCES

1. William Wahl, *A Practical Guide for the Gold and Silver Electroplater and the Galvanoplastic Operator*, Henry Carey Baird & Co., Philadelphia, 1883.
2. John Hunter III, "Minimizing Waste by Source Segregation and Inventory Control," *Technical Strategies for Hazardous Waste Prevention and Control Seminar*, Government Institutes, Inc., Washington, D.C., 1987.
3. Jerome Kohl et al. *Managing and Recycling Solvents in the Furniture Industry*, North Carolina Board of Science and Technology, Raleigh, N.C., 1986.
4. Susan Boyle and Raghu Raghawan, "Hazardous Waste Source Reduction Case Studies," *Proceedings of the CMA Waste Minimization Workshop*, vol. 1, Chemical Manufacturing Association, Washington, D.C., 1987.
5. Jerome Kohl, *Hazardous Waste Minimization: The New RCRA Initiative*, Industrial Extension Service, North Carolina State University, Raleigh, 1988.
6. R. A. Mead, "Waste Reduction at a Paint Plant—An OE Approach," *Proceedings of the Hazardous Waste Reduction Audit Workshop*, New Jersey Department of Environmental Protection, Trenton, N.J., 1987, pp. 156–163.
7. Gregory Lorton, "Waste Minimization in the Paint and Allied Products Industry," *J. Air Pollution Control Association*, 38:4, 1988.
8. Ryan Delcambre, "Waste Reduction: Program Practice and Product at Dow Chemical," *Waste Minimization and Recycling Report*, 1(1):7, 1986.
9. *Governor's Award for Excellence in Waste Management—1987*, Governor's Waste Management Board, Raleigh, N.C., 1988.
10. Paul Dadak, "Waste Minimization: The Hewlett Packard Experience," *Waste Minimization Manual*, Government Institutes, Inc., Washington, D.C., 1987.
11. Stephen Mims, "Glycol II—Absorber Water Upgrade Project," *Proceedings of the CMA Waste Minimization Workshop, vol. 1*, Chemical Manufacturers Association, Washington, D.C., 1987.
12. Steven Rice, "Waste Reduction in the R & D Industry," *Proceedings of the Hazardous Waste Reduction Audit Workshop*, New Jersey Department of Environmental Protection, Trenton, N.J., 1987, pp. 181–190.
13. *Governor's Award for Excellence in Waste Management—1986*, Governor's Waste Management Board, Raleigh, N.C., 1987.
14. David J. Sarokin et al., *Cutting Chemical Wastes: What 29 Organic Chemical Plants Are Doing to Reduce Hazardous Wastes*, Inform, Inc., New York, 1985.
15. Rupert Bullard, John Rushing, and Roy Carawan, "A Dairy Processor Does It," *Proceedings of the Conference Waste Reduction Pollution Prevention: Progress and Prospects within North Carolina*, Pollution Prevention Program, North Carolina Department of Natural Resources and Community Development, Raleigh, N.C., 1988, pp. 29. 1–29.6.
16. David Adkins, Gary Hunt, and Roger Schecter, *Pollution Prevention Challenge Grants: Project Summaries*, Pollution Prevention Program, North Carolina Department of Natural Resources and Community Development, Raleigh, N.C., 1988.
17. Donald Huisingh, *Profits of Pollution Prevention: A Compendium of North Carolina Case Studies*, North Carolina Board of Science and Technology, Raleigh, N.C., 1985.
18. Case Study Files from the Pollution Prevention Program, Division of Environmental Management, North Carolina Department of Natural Resources and Community Development, Raleigh, N.C.
19. Herbert Skovronek, "Paint Application: Waste Management," *Proceedings of the Hazardous Waste Reduction Audit Workshop*, New Jersey Department of Environmental Protection, Trenton, N.J., 1987, pp. 164–172.
20. Charles Parmele, Robert Fox, and Richard Standifer, "Waste Minimization Through Effective Chemical Engineering," *Proceedings of the Conference on Performance and Costs of Alternatives to Land Disposal of Hazardous Waste*, Air Pollution Control Association, Pittsburgh, Pa., 1986, pp. 115–123.
21. Carl Fromm and Sunivas Budarajer, "Reducing Equipment Cleaning Wastes," *Chemical Engineering*, 95(10):117, 1988.
22. Thomas Kalenowski and Mary Ann Keon, *Waste Generation and Disposition Practices and Currently Applied Waste Minimization Techniques within the Semiconduc-*

tor Industry, State of California Department of Health Services, Sacramento, Ca., 1987.

23. Roy Carawan, James Chambers, and Robert Zall, *Management Control of Water Use and Wastes in Food Processing*, AM-18L, Agricultural Extension Service, North Carolina State University, Raleigh, N.C., 1979.

24. Robert Shober, "Water Conservation and Waste Load Reduction in Food Processing Facilities," *1988 Food Processing Waste Conference Proceedings*, Georgia Tech Research Institute, Georgia Institute of Technology, Atlanta, Ga., 1988.

25. U.S. Environmental Protection Agency, *Waste Minimization Issues & Options: Vol. II*, EPA/530-SW-86-042, Office of Solid Waste and Emergency Response, Washington, D.C., 1986.

26. R. Keith Mobley, "Turning Maintenance Dollars into Bottom-Line Profits," *CPI Equipment Reporter*, May/June 1988, p. 21.

27. John Hunter III, "Life Cycle Approach to Effective Waste Minimization," paper presented at *Conference on Engineering to Minimize the Generation of Hazardous Waste*, Engineering Foundation, Hinniker, N.H., 1987.

28. Bill Pitchford, "Process Modification: One Alternative to Chlorinated Solvents," *Focus Waste Minimization*, North Carolina Department of Human Resources, Issue IV, 1987, pp. 5–6.

29. George Makrauer, "Water-Based Inks in Flexographic Printing—High Slip Polyethylene Film," *Proceedings of the Conference Waste Reduction—Pollution Prevention: Progress and Prospects within North Carolina*, Pollution Prevention Program, North Carolina Department of Natural Resources and Community Development, Raleigh, N.C., 1988.

30. Gary Hunt and Roger Schecter, *Accomplishments of North Carolina Industries: Case Summaries*, Pollution Prevention Program, North Carolina Department of Natural Resources and Community Development, Raleigh, N.C., 1987.

31. *Governor's Conference on Pollution Prevention Pays Proceedings*, Governor's Safe Growth Team, Nashville, 1986.

32. T. E. Higgins, *Industrial Processes to Reduce Generation of Hazardous Waste at DOD Facilities, Phase I Report Evaluation of 40 Case Studies*, Department of Defense Environmental Leadership Project Office, Washington, D.C., and U.S. Army Corps of Engineers, Huntsville, Ala., 1985.

33. Jerome Kohl, Phillip Moses, and Brooke Triplett, *Managing and Recycling Solvents: North Carolina Practices, Facilities, and Regulations*, North Carolina Board of Science and Technology, Raleigh, N.C., 1984.

34. Jerome Kohl, Jeremy Pearson, and Brooke Triplett, "Reducing Hazardous Waste Generation with Examples from the Electroplating Industry," *Waste Management Advisory Note*, vol. 22, Solid and Hazardous Waste Management Branch, North Carolina Department of Human Resources, Raleigh, N.C., 1986.

35. J. Edward Smith, *Evaluation of a Teflon-Based Ultraviolet Light System on the Disinfection of Water in a Textile Air Washer*, Pollution Prevention Program, North Carolina Department of Natural Resources and Community Development, Raleigh, N.C., 1987.

36. 3M Company, "Riker Innovation Meets Air Regulation," *Ideas: A Compendium of 3P Success Stories*, 3M Company, Saint Paul, Minn.

37. Alice Johnson, "Experiences in Getting Rid of Solvent-Based Degreasing in a Diesel Engine Remanufacturing Plant," *Proceedings of the Conference Waste Reduction—Pollution Prevention: Progress and Prospects within North Carolina*, Pollution Prevention Program, North Carolina Department of Natural Resources and Community Development, Raleigh, N.C., 1988, pp. 15.1–15.5.

38. Gary Hunt, "Waste Reduction in the Metal Finishing Industry," *J. Air Pollution Control Association*, 38(5):672, 1988.

39. W. Beck, "Waste Minimization—A Plant Approach to Getting Started," *CMA Waste Minimization Workshop Proceedings*, vol. I, Chemical Manufacturers Association, Washington, D.C., 1987.

40. Jerome Kohl and Brooke Triplett, *Managing and Minimizing Hazardous Waste*

Metal Sludges: North Carolina Case Studies, Services and Regulations, Governor's Waste Management Board and the Hazardous Waste Management Branch, North Carolina Department of Human Resources, Raleigh, N.C., 1984.

41. Monica E. Campbell and William M. Glenn, *Profit from Pollution Prevention, A Guide to Industrial Waste Reduction and Recycling*, Pollution Probe Foundation, Toronto, 1982.

42. Versar, Inc., and Jacobs Engineering Group, *Waste Minimization Issues and Options*, vol. III, EPA 530-SW-86-043, Office of Solid Waste, U.S. Environmental Protection Agency, Washington, D.C., 1986.

43. Carl H. Fromm and Michael S. Callahan, "Waste Reduction Audit Procedure—A Methodology for Identification, Assessment, and Screening of Waste Minimization Options," in *Proceedings of the National Conference on Hazardous Wastes and Hazardous Materials*, Hazardous Materials Control Research Institute, Silver Springs, Md., 1986.

44. *U.S. EPA Summary Report—Control and Treatment Technology for the Metal Finishing Industry—In-Plant Changes*, EPA-625/8-82-008, U.S. Environmental Protection Agency, Cincinnati, 1982.

45. Davis Lewis, "Waste Minimization in the Pesticide Formulation Industry," *J. Air Pollution Control Association*, 38(10):1293, 1988.

46. James Burke, *Connections*, Little, Brown & Co., Boston, 1978.

47. James Waynick, Roy Carawan, and Fred Tarver, "A Breaded Foods Processor Does It Too!" *Proceedings of the Conference Waste Reduction—Pollution Prevention: Progress and Prospects within North Carolina*, Pollution Prevention Program, North Carolina Department of Natural Resources and Community Development, Raleigh, N.C., 1988, pp. 30.1–30.9.

48. Jerome Kohl and John Currier, *Managing Waste Oils*, Industrial Extension Service, North Carolina State University, Raleigh, N.C., 1987.

49. Thomas J. Nunno and Mark Arienti, "Waste Minimization Case Studies for Solvents and Metals Waste Streams," presented at the *Hazardous Waste Disposal Conference*, U.S. Environmental Protection Agency, Cincinnati, 1986.

50. Resource Integration Systems, Inc., J. L. Richards & Associates Limited, Ontario Research Foundation, and The Proctor and Redfern Group, *Technical Manual: Waste Abatement, Reuse, Recycle, and Reduction Opportunities in Industry*, Environment Canada, Toronto.

51. Catherine Patriquen, "Beneficial Use of Wastes within Specialty Chemicals Manufacturing," *Proceedings of the CMA Waste Minimization Workshop*, vol. II, Chemical Manufacturers Association, Washington, D.C., 1987.

52. Brent Smith, *Identification and Reduction of Pollution Sources in Textile Wet Processing*, Pollution Prevention Program, North Carolina Department of Natural Resources and Community Development, Raleigh, N.C., 1986.

53. James Nesmith III, "Marketing Industrial Waste: A Generator's Perspective," *Proceedings of the Fourth National Conference on Waste Exchange*, Southeast Waste Exchange, Charlotte, N.C., 1987.

54. Edward Shields, "Transforming Wastes into Products," *Proceedings of the Source Reduction of Hazardous Waste Seminar*, New Jersey Department of Environmental Protection, Trenton, N.J., 1985.

55. Travis A. Turberville, "Market Development for Fumed Amorphous Silica and Ladle Rakeout," *Proceedings of the Conference Pollution Prevention Pays: Waste Reduction in Alabama*, University of Alabama Center for Urban Affairs, Birmingham, 1985.

56. Kenneth E. Noll et al., *Recovery, Recycle and Reuse of Industrial Waste*, Lewis Publishers, Chelsea, Mich., 1985.

57. Harry Freeman (ed.), *Standard Handbook of Hazardous Waste Treatment and Disposal*, McGraw-Hill Book Co., New York, 1988.

58. Richard Dowing, "Marketing Waste Byproducts as Reusable Feedstocks (Part 1)," *Waste Minimization and Recycling Report*, Issue 22, 1988, pp. 7–9.

59. David Jacobs, "Brokering of Industrial By-Products," *Proceedings of the Fourth Na-*

tional Conference on Waste Exchange, Southeast Waste Exchange, Charlotte, N.C., 1987.

60. Bill Stough, "Waste Exchanges: Catalysts for Innovation in Waste Minimization," *J. Air Pollution Control Association*, 38(6):744, 1988.

61. Terry Schurter, "Waste Marketing: A Case Study from the Electroplating Industry," *Proceedings of the Conference Waste Reduction—Pollution Prevention: Progress and Prospects within North Carolina*, Pollution Prevention Program, North Carolina Department of Natural Resources and Community Development, Raleigh, N.C., 1988, pp. 33.1–33.3.

62. *National Trade and Professional Association of the U.S.—1986*, Columbia Books, Washington, D.C., 1986.

63. *Solvent Waste Reduction Alternatives Seminar*, U.S. Environmental Protection Agency, Chicago, 1988.

Waste Minimization in Industry

Implementing Waste Minimization Programs in Industry

Gregory J. Hollod

Conoco, Inc
Houston, Texas

William B. Beck

E. I. duPont
Houston, Texas

4.1 Getting Started

The first step for the individual in the company or the manufacturing site who is faced with the challenge of establishing and implementing a waste minimization program is to review the incentives and barriers for getting the program started. Then the essential elements of the program must be defined.

The four basic incentives which will motivate corporate senior management or a plant manager to establish and drive a waste minimization program are (1) the real cost and economics associated with generating and managing wastes; (2) corporate policies, procedures, and waste reduction goals; (3) the understanding that waste minimization will improve a company's environmental position by reducing risk, improving public support, and better utilizing natural resources; and (4) legal requirements, such as routine audits by the state, required waste minimization plans, and demonstrated progress on waste minimization as a prerequisite for obtaining a new or modified permit and waste end tax.

Little is known or appreciated about the true costs associated with generating hazardous waste. If this information were available, it would heighten the awareness of management to the need for different waste management programs and it would support many capital projects aimed at reducing waste.

The four basic costs incurred when a company generates waste arise from (1) underutilizing the value of the raw materials; (2) general management costs associated with moving the waste around the site, storage, keeping track of waste records, and shipping; (3) disposing of the waste; and (4) the associated costs with third-party liabilities if the waste is improperly disposed of. We know or can calculate the first three costs, but it has been very difficult to obtain any real values for the costs associated with remediating improperly disposed materials and for Superfund costs. Attempting to determine the true management costs associated with generating wastes at a particular company, either at a plant site or from a product line, will give managers the opportunity to truly realize waste management costs, and waste minimization will become much more attractive.

The second type of incentive is to develop corporate policies and procedures and set goals for the organization to "march" to. At DuPont, we established a policy in 1980 which stated that we intend "to minimize the generation of waste to the extent that is technically and economically feasible." This policy provides a catalyst to emphasize waste minimization on a continuing basis and make it a part of our culture.

DuPont has been aware of waste minimization and actively involved in it since the company was founded in 1802. In the early days, managers and operators alike worked continually to make the manufacturing of black gun powder cheaper and safer. It was recognized that process waste was hazardous in a very immediate sense, and it was to everyone's advantage to keep waste to a minimum. Today, instead of minimizing the losses of sulphur, nitrate, and charcoal from the manufacturing of gun powder, we are focusing our attention on minimizing the losses of chlorine, benzene, and precious metal catalyst from the manufacturing of Kevlar, nylon, and other products.

In addition to policies, many companies have actually established numerical goals for waste reduction. For instance, at DuPont, the goal is that waste will be reduced by 35 percent by 1990 when compared to 1982 values. Chevron Corporation has committed to a 60 percent reduction for worldwide operations over the next four years. These goals will provide the types of incentives that plant managers and chemical engineers are accustomed to striving for and achieving.

Progress is being made in waste minimization. For instance, in the chemical industry hazardous waste generated between 1981 and 1986 decreased by 10 percent, according to a recent Chemical Manufacturer's Association survey. This example, in addition to documented waste reductions at the Department of Defense, General Dynamics, General Electric, Polaroid, and smaller companies like Rexham Corporation in North Carolina and Servtron Inc. in Minnesota, demonstrates that waste reduction programs are underway and goals are being achieved.

It is prudent to be aware that the dangers inherent in establishing waste minimization goals are the same dangers we have with regulations in general, that is, we may begin to play a "numbers game" or develop a mind-set that says if we satisfy the regulation or if we achieve the number, we've won the game. If we have learned anything about environmental policy, it's that playing the regulatory numbers game is not conducive to innovation. Many of us don't feel obligated to become creative as long as we make the numbers. We must not lose sight of the longer-term goal (that is, to eliminate waste and emissions) by allowing ourselves to be placated by the short-term successes. If we play the numbers game instead of going after real technical solutions and innovations, we will always be "playing catch-up."

Waste minimization has often been referred to as a double-edged sword. Although it may require some capital to begin a waste reduction program, reducing waste will reduce costs and improve revenue and at the same time improve the overall environmental position. Although difficult to measure, there will also be greater community and public support for plant activities. Waste reduction is tangible evidence of a company's or plant site's commitment to clean up the environment and to keep it clean.

The final incentive to initiate waste minimization programs is a review and consideration of legal requirements. There are many bills and pieces of draft legislation now in existence which require waste minimization. They range from multimedia data collection to efficiency standards for waste treatment. By going the "legal" route, there is a risk of having less support from or even having resentment from the line organization about participating in the program. Many times imagination, creativity, and willingness to participate can be stifled by regulations. However, as we have seen in the past, this country tends to go more toward the legal and legislative route than toward any other type of program. In the future we will witness significant changes in the regulatory agenda, and waste minimization will be pulled into the environmental regulatory structure like other environmental issues of the past.

4.2 Barriers

There are often many good and valid reasons why waste minimization programs may not be moving at the rate management and the public would like. There are several potential barriers, including a shortage of capital, lack of technical personnel, competing priorities, attitudes about managing waste, lack of senior management support, and finally, lack of full appreciation for the need to minimize waste.

One of the difficulties that any project or program encounters in industry is the competition for capital. There is only so much money that can be spent or utilized by the operating departments. This creates a competitive situation where the projects with the highest rate of return will be funded. Since the true economics and cost associated with waste management as discussed above are not known, it is difficult to value properly an investment made for reducing waste. In addition, many waste minimization ideas are at first glance an expensive solution. Good understanding of how and why wastes are produced is often lacking. Improving this knowledge will lower the costs of the solutions.

There are options which a company can use to overcome this competition for capital for waste minimization programs. Some companies provide a pool of corporate money which can be used for waste minimization projects in a way that is similar to the funding that was used in the 1970s for energy conservation programs. Companies can set aside money in a corporate pool, and any project associated with reducing waste will be approved and funded. However, management must recognize that many capital projects initiated in the 1970s could not be justified with the current price of oil. Similarly, the price for managing waste may eventually stabilize and perhaps even be lowered as new technologies for treatment and new remediation techniques are developed and more commercial treatment systems come on line.

The lack of technical personnel can restrict a waste minimization program. Many companies today, in an effort to become more cost-competitive, have cut their technical staffs. In addition, there is really no technical pool of resources available in the consulting area or outside of companies in general that can be tapped or be used to provide this technical expertise. There is a dearth of process-related, technically trained individuals who appreciate the need for reducing waste under the current environmental regulatory regime.

Lack of senior management support is a factor that will doom a waste minimization program from the start. Many managers, in addition to their standard business functions, have become occupied with other priorities in the environmental area such as land bans, right-to-know, and occupational health considerations. Waste minimization is competing with other environmental priorities. For a program to succeed, management must be convinced that waste minimization deserves priority and should be part of the "daily diet" for the line organization—and not just another environmental headache left to the site's environmental coordinator.

At DuPont we recognize that for waste minimization to be successful, it must be part of the line organization. The designers, builders, and operators are the ones who must be educated to make the changes necessary for reducing wastes. The environmental community in a company should act like cheerleaders, raising the flag of awareness and providing regulatory advice to encourage the line organization to reduce waste.

To effectuate the move of waste minimization concern into the line organization, DuPont held its second corporate waste minimization conference on November 13, 1988. The emphasis for the 300 attendees at Waste Minimization II was involvement by line organization. Plant managers, process engineers, and operators participated in panel discussions and gave a significant percentage of the 30 papers presented at the conference.

4.3 Essential Components

Once the incentives and barriers have been addressed for the specific company and senior level management supports the program, the three essential strategies necessary to make a successful waste minimization program are (1) to designate members who will make up a well-defined organization created to follow the program, (2) to define a target of waste to reduce, and (3) to develop a method for tracking the performance and progress of the program. All three are necessary, but each component has its own separate functions and limitations.

4.3.1 The Site Organization

The role of the site organization is characterized in Fig. 4.1. The most important functions are to provide awareness and to identify what needs to be done, which includes raising the economic flag so as to en-

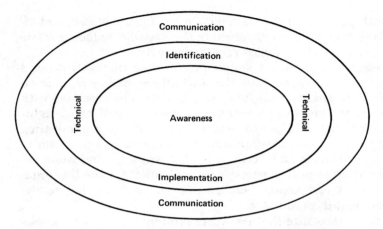

Figure 4.1   The role of a waste minimization organization.

sure prompt action. The second most important function of the organization is to provide communication, both inside and out of the company. Inside there needs to be a cross-fertilization of ideas, what one group is doing, what another department should know and how one product has become more competitive because of its improved waste management position. Outside of the company, communications are needed to demonstrate that progress has in fact been made for reducing waste. Outside communications can also be effective in bolstering inside momentum, especially if these outside communications compare the company favorably to its competition, showing that the company is better because of its environmental efforts.

Table 4.1 provides a detailed list of definitive functions for the site organization, beginning with defining the objectives and reviewing the program with site management and going all the way through auditing the program and recognizing achievements with awards. This

TABLE 4.1   Functions of Waste Minimization Site Organization

Define objectives	Coordinate program participation
Review objectives with site management	Set up accounting system
Communicate objectives to site	Upgrade projects
Get support from generators	Schedule reviews
Get representation from areas	Conduct audits
Provide ongoing awareness and training	Summarize site progress
Provide resources	Recognize achievements
Catalyze program participation	

is the first environmental program to come along that does not need a lawyer to interpret the appropriate program or meaning of the program. The site committee should have an easier task communicating objectives to and obtaining support from the line organization than it would have with other recent environmental programs, such as the Resource Conservation and Recovery Act (RCRA), the Clean Water Act, and the Clean Air Act. The objective of the waste minimization organization is to define a functional system that works which includes representation and support from all the vital components of the manufacturing process. In other words, since the line organization has the knowledge of the process and controls the capital, it is the line organization that is ultimately responsible for reducing waste.

To ensure the success of the program, the waste minimization organization should establish an accounting system which allows meaningful reviews and audits to be conducted. As the program evolves, it is the responsibility of the central organization to recognize and review successful efforts. This recognition will help perpetuate the program and raise people's awareness level of it.

4.3.2   Target Definition

The current lack of any national regulatory program allows each manufacturing facility the flexibility of selecting targets based on individual business needs. This is the ideal situation in which truly the intent of minimization is to not only protect the environment but also to make businesses more competitive. Wastes that are either large in volume or expensive to manage and to dispose of should be the highest priority targets for reduction. Table 4.2 summarizes the parameters that should be considered when selecting a target.

Many companies have selected targets of wastes to reduce that are toxic or RCRA-regulated. At DuPont, a uniform target of waste to reduce was defined in 1984. We called it "DuPont tabulated waste" (Table 4.3). The target characterizes the wastes in accordance with the manner in which they are produced or managed, which is obviously much broader and more encompassing than RCRA-defined waste per

TABLE 4.2   Bases for Definition of a Waste Reduction Target

General site needs	Waste with highest associated costs
Corporate goals	Hazardous components
Business objectives	Future products
Volume of waste	

TABLE 4.3   General Description of DuPont Tabulated Waste

Disposal/Use of Waste

"Solids" treated or disposed of off site (including wastewaters)

"Solids" treated or disposed of on site (except for wastewaters, see below)

Barged to sea

Injected into deep wells

Total influent to on-site wastewater treatment facilities

Total influent to publicly owned treatment works (POTW)

Used for fuel

se. This means of characterizing wastes has particular advantages in that it provides a consistency across the company, so that all plants can work toward a common goal and the performance of all can be compared equitably. As business needs change and our waste reduction program evolves, we may need to include still other wastes or modify the definition.

Once waste has been targeted and a program for reduction has been implemented, it then becomes important to track performance. Tracking the progress of waste reduction is no different from tracking any other production variable. However, to design the tracking function properly one must decide beforehand if the data are to be used exclusively for internal corporate needs for tracking the program, shared with the technical resources in the company, or shared with external groups. Clearly, the information needs of a business or technical manager making a decision are different from the informational needs of a regulatory or public interest group.

4.3.3 Tracking

A tracking system can be used to identify waste reduction opportunities and provide essential feedback to successfully guide future efforts. Considerations to volume of waste, cost of handling, regulatory impact, product life cycle, marketing opportunities, and basic manufacturing processes can be factored into an algorithmic function and coupled with a tracking system to identify opportunities. The identification process would allow a business team to maximize the technical and capital resources available and direct them to the part of the business that needs them the most to improve its overall performance.

A dynamic data base using Datatrieve, a Digital Equipment Co. product, has been developed to track the waste targeted by DuPont.

TABLE 4.4   Sample of Typical Computer Printout for Waste Tracking

Production area	Waste description	Hazardous classification	Quantity generated lb/yr (in Thousands)	Management disposal cost $/yr (in Thousands)	Minimization method
VI023	Organic acid	Flammable	2	5.5	Recycle
NR126	Polymers	Caustic	50	25	Sale
GA462	Spent catalyst	Acidic	10	42	Reuse
ME621	Lab solvent	Ignitable	0.5	1	Fuel
BU215	Acid catalyst	Corrosive	40	16	Administrative control

Datatrieve is an interactive language that organizes information into collections of interrelated data or data bases. Its use requires a VAX/VMS mainframe operating system. Once the data have been entered into the data base, a variety of reporting functions are available.

The tracking function or recordkeeping at a minimum should record and identify the generator or "owner" of the waste, associated volumes, cost of managing the waste, and the waste reduction method being used to reduce that particular wastestream. Table 4.4 shows a typical printout from a computerized tracking program that has been used at DuPont.

One reporting function that would be of particular interest to any program is the tracking of the most successful or most often used waste minimization techniques. Table 4.5 lists the validation codes for the typical waste minimization techniques that have been used at DuPont. This information can be used by business managers and technical managers to inform manufacturing facilities in different parts of the country as to what might be the most successful waste minimization techniques to apply.

4.4   Action Plan

Table 4.6 gives a list of the various elements in a waste minimization program, with the corresponding responsibilities and the timing for the various actions to be completed. The waste minimization organization should be responsible for tracking this action plan using audits and monthly reviews. The organization must apply various techniques

TABLE 4.5   Validation Codes for Typical Waste Minimization Techniques

10:	Process change
11:	Modify operating procedure
12:	Advanced process control
13:	Substituted chemicals
14:	Use higher quality materials
20:	Recycle
21:	Direct use in the process
22:	Direct use in another process
23:	Regeneration for reuse
24:	Use as a fuel
25:	Sale
30:	Improve waste treatment
31:	Waste filtration
32:	Waste decantation
33:	On-line treatment
40:	Administrative controls
41:	Minimizing washdown
42:	Reduce cleaning frequency
43:	Longer turnaround time
44:	Improved spill control
45:	Separate hazardous from nonhazardous
46:	Discontinue manufacture

TABLE 4.6   Sample Action Plan for Waste Minimization Program (Blank)

Elements	Responsibility	Timing
Employee awareness		
Identify all waste and owners		
Update waste standards		
Technical review of procedures		
Cost system for treatment		
Computer tracking		
Monthly review		
Auditing		
Historical data		
Organize data		
Report		
Recognize achievements		

to maintain the momentum of the program; these techniques can be divided into short-term and long-term initiatives.

There are many short-term solutions that can be implemented now for reducing waste. Foremost among these is the establishment of training programs.

At DuPont a unique review program has been developed which provides a series of methodologies and techniques to evaluate present waste management systems. This review is called the Minimization Survey and it is a tool to assist plant personnel with establishing a program for identifying opportunities. The concept of the survey is to provide a useful working document to those closest to the process. Instead of bringing in outsiders who might know waste minimization but don't know the process, we have taken the approach of educating those most familiar with the process to the needs associated with a waste minimization program. For instance, general programmatic items such as identifying the waste, determining the disposal practices and the cost, educating employees about new material substitution, evaluating current practices, and establishing plant goals all have been outlined and described in detail in this survey. Once these ideas have been reviewed alongside considerations for changes in material handling and available technology, a waste management analysis can then be conducted at a plant site.

In the Waste Minimization Survey, the first item to be reviewed is the very basic operations of the plant manufacturing process (Fig. 4.2). Building upon this information the process should then be defined in more detail (Fig. 4.3). Tracking the information outlined in the column headers of Table 4.4 for the important wastestreams or targeted waste will generate data which then can be utilized in implementing the waste reduction analysis process outlined in Fig. 4.4. This waste reduction analysis process is unique in that it outlines a step-by-step approach for identifying and developing a path toward long-term solutions.

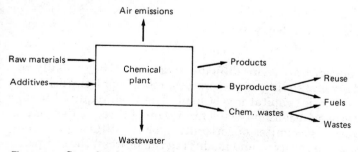

Figure 4.2 Superficial waste management analysis.

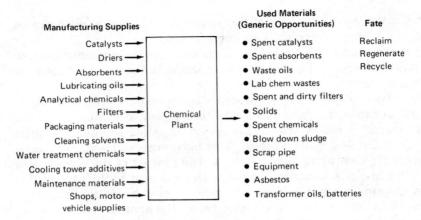

Figure 4.3 In-depth waste management analysis.

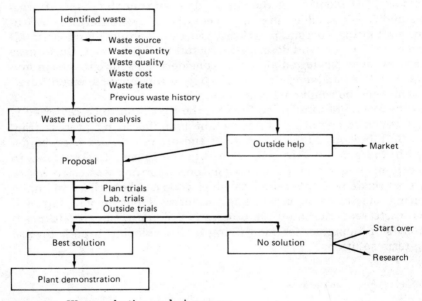

Figure 4.4 Waste reduction analysis process.

Long-term solutions are more difficult to achieve. They usually require a better understanding of the process, more sophisticated technology, and substantial capital investment. The time required to implement long-term solutions may be a minimum of three to five years. As the organization recognizes opportunities for substituting raw materials, changing the catalyst, and modifying the process, the company will steadily move toward long-term improvements in the waste management program.

Reducing waste to remain competitive has been an important ingredient for successful business in the past. In the future it will be absolutely essential. Although industry has made substantial progress in responding to economic and regulatory incentives, it must recognize that a more intensive effort will be needed in the future.

Waste Minimization Assessments

**Deborah Hanlon
and
Carl Fromm**

*Jacobs Engineering Group
251 South Lake Avenue
Pasadena, California*

Waste minimization is a policy specifically mandated by the U.S. Congress in the 1984 Hazardous and Solid Wastes Amendments to the Resource Conservation and Recovery Act (RCRA). This mandate, coupled with other RCRA provisions that have led to unprecedented increases in the costs of waste management, have heightened general interest in waste minimization. A strong contributing factor has been a desire on the part of generators to reduce their environmental impairment liabilities under the provisions of the Comprehensive Environmental Response, Compensation, and Liabilities Act (CERCLA, or "Superfund"). Because of these increasing costs and liability exposure, waste minimization has become more and more attractive economically.

The RCRA regulations require that generators of hazardous waste "have a program in place to reduce the volume and toxicity of waste generated to the extent that is economically practical." A waste minimization program is an organized, comprehensive, and continual effort to systematically reduce waste generation. Generally, a program is established for an organization as a whole. Its components may include specific waste minimization projects and may use waste minimization assessments as a tool for determining where and how waste can be reduced. A waste minimization program should reflect the goals and policies for waste minimization set by the organization's management. Also, the program should be an ongoing effort and should strive to make waste minimization part of the company's operating philosophy. While the main goal of a waste minimization program is to reduce or eliminate waste, it may also bring about an improvement in a company's production efficiency.

The Waste Minimization Opportunity (WMO) assessment is a structured project activity designed to help identify, screen, and analyze options to reduce the volume and toxicity of wastestream(s) or emission(s) leaving the target process or operation. The procedure itself is merely a management tool created to identify and characterize waste minimization options in a systematic and comprehensive fashion. The assessment (or "waste audit," as some people call it) is regarded as a principal component of the waste minimization program generically represented in Fig. 5.1. Readers should note that WMO assessments are the key mechanism for generating candidate projects, which are then subjected to screening, funding, implementation, and performance evaluation.

Many people mistakenly confuse waste audits with environmental audits. Environmental audits seek to characterize regulatory compliance status. Waste audits or assessments are performed solely to determine ways in which to minimize hazardous wastes or emissions via source reduction and recycling. The procedure described in this chapter is virtually identical to the one recommended by the U.S. Environmental Protection Agency (EPA) in its manual.[32] The procedure was developed based on the initial concept and was then tested and recti-

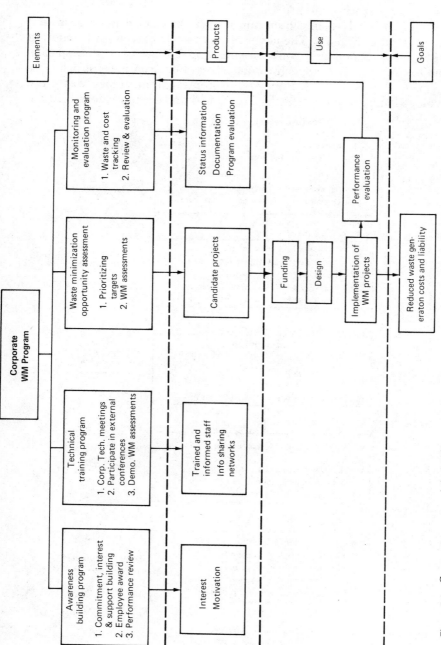

Figure 5.1 Corporate waste minimization program.

fied through a series of EPA-sponsored pilot assessments carried out at various operating facilities.

The steps involved in conducting a waste minimization assessment are outlined in Fig. 5.2. The assessment consists of a careful review of a plant's operations and wastestreams and the selection a of specific area to assess. After a specific wastestream or area is established as the focus, a number of options with potential to minimize waste are developed and screened. Then the technical and economic feasibility of

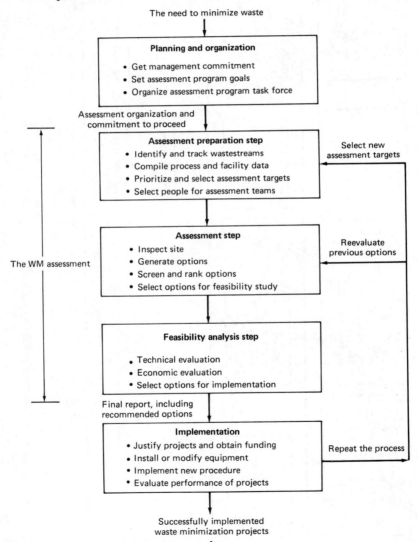

Figure 5.2 Waste minimization procedure.

the selected options are evaluated, and finally, the most promising options are selected for implementation.

Because a comprehensive waste minimization program affects many functional groups within a company, the program needs to bring these different groups together to reduce wastes. The formality of the program depends upon the size and complexity of the organization and its waste problems. The program structure must be flexible enough to accommodate unforeseen changes. The developmental activities of a waste minimization program include

- Getting management commitment
- Setting waste minimization goals
- Staffing the program task force

5.1 Program Planning and Organization

The developmental phase of the waste minimization program centers on organization and planning (see Fig. 5.3).

5.1.1 Getting management commitment

The management of a company will support a waste minimization program if it is convinced that the benefits will outweigh the costs. The potential benefits include economic advantages, compliance with reg-

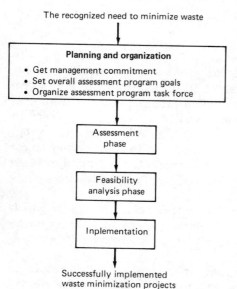

Figure 5.3 Planning and organization phase.

ulations, reduction in liabilities associated with the generation of wastes, improved public image, and reduced environmental impact.

The objectives of a waste minimization program are best conveyed to employees through a formal policy statement of management directive. Upper management is responsible for establishing a formal commitment throughout all divisions of the organization. The company's environmental affairs officer is responsible for advising management of the importance of waste minimization and the need for this formal commitment. An example of a formal policy statement follows:

> [A major chemical company] is committed to continue excellence, leadership, and stewardship in protecting the environment. Environmental protection is a primary management responsibility, as well as the responsibility of every employee.
>
> In keeping with this policy, our objective as a company is to reduce waste and achieve minimal adverse impact on the air, water, and land through excellence in environmental control.
>
> The Environmental Guidelines include the following points:
> Environmental protection is a line responsibility and an important measure of employee performance. In addition, every employee is responsible for environmental protection in the same manner he or she is for safety.
>
> Minimizing or eliminating the generation of waste has been and continues to be a prime consideration in research, process design, and plant operations and is viewed by management like safety, yield, and loss prevention.
>
> Reuse and recycling of materials has been and will continue to be given first consideration prior to classification and disposal of waste.

5.1.2   Getting employees involved

Although management commitment and direction are fundamental to the success of a waste minimization program, commitment throughout an organization is necessary in order to resolve conflicts and to remove barriers to the waste minimization program. Waste minimization should become part of the corporate culture, permeating all strata and activities. Bonuses, awards, plaques, and other forms of recognition are often used to provide motivation and to boost employee cooperation and participation. In some companies, meeting the waste minimization goals is used as a measure for evaluating the job performance of managers and employees.

Technical staff will benefit from a technical training program which emphasizes waste minimization. The elements may include corporate technical meetings, participation in external conferences, demonstration audits, and workshops.

Any waste minimization program needs one or more people to champion the cause. These "cause champions" help overcome the inertia present when changes to an existing operation are proposed. They lead the waste minimization program, either formally or informally. An environmental engineer, production manager, or plant process engineer may be a good candidate for this role. Regardless of who takes the lead, the cause champion must be given enough authority to carry out the program effectively.

5.1.3 The program task force

The waste minimization program will affect a number of groups within a company. For this reason, a program task force should be assembled. This task force should include members of any group or department in the company that has a significant interest in the outcome of the program.

A waste minimization program task force typically is responsible for

- Getting a commitment and a statement of policy from management
- Establishing overall waste minimization program goals
- Establishing a monitoring system
- Propagating waste minimization culture throughout the organization
- Prioritizing the wastestreams or facility areas for assessment
- Selecting assessment teams
- Conducting (or supervising) assessment
- Conducting (or monitoring) technical and economic feasibility analyses of favorable options
- Selecting and justifying feasible options for implementation
- Obtaining funding and establishing a schedule for implementation
- Monitoring (and/or directing) implementation progress
- Monitoring performance of the option, once it is operating

Figure 5.4 depicts an example worksheet used to identify potential departments and personnel to consider when selecting assessment teams. While the waste minimization program task force is concerned with the whole plant, the focus of each of the assessment teams is more specific, concentrating on a particular area of the plant. Each team should include people with direct responsibility and knowledge of the particular wastestream or area of the plant.

In addition to the internal staff, consider using outside people to help the task force and the assessment teams, especially in the assess-

Firm_____	Waste minimization assessment	Prepared by _____
Site _____	Proc. unit oper. _____	Checked by _____
Date _____	Proj. no. _____	Sheet $\underline{1}$ of $\underline{1}$ Page ___ of ___

Worksheet		Assessment team makeup

Function department	Name	Location/ telephone /#	Manhours required	Duties		
				Lead	Support	Review
Assessment team						
Leader						
Site coordinator						
Operations						
Engineering						
Maintenance						
Scheduling						
Material control						
Procurement						
Shipping/receiving						
Facilities						
Quality control						
Environmental						
Accounting						
Personnel						
R & D						
Legal						
Management						
Contractor/consultant						
Safety						

Figure 5.4   Assessment team makeup.

ment and implementation phases. They may be consultants or experts from a different facility of the same company. In large multidivision companies, a centralized staff of experts at the corporate headquarters may be available. One or more "outsiders" can bring in new ideas and provide an objective viewpoint. An outsider also is more likely to counteract bias brought about by "inbreeding" or the "sacred cow" syndrome; such bias may be evident when an old process area, rich in history, undergoes an assessment.

Production operators and line employees must not be overlooked as a source of waste minimization suggestions, since they possess first-hand knowledge and experience with the process they work on. Their assistance is especially useful in assessing operational or procedural changes or equipment modifications that affect the way they do their work. "Quality circles" have been instituted by many companies, particularly in manufacturing industries, to improve product quality and production efficiency. These quality circles consist of meetings of workers and supervisors at which improvements are proposed and evaluated. Quality circles are beneficial in that they involve the production people who are closely associated with the operations and foster participation and commitment to improvement. Several large companies that have quality circles have used them as a means of soliciting successful suggestions for waste minimization.

The formality or informality of a company's waste minimization program will depend on the nature of the company. The program in a highly structured company will probably develop to be quite formal, in contrast to the usually more informal waste minimization program in a small company or a company in a dynamic industry in which the organizational structure changes frequently.

In a small company, several people at most will be all that are required to implement a waste minimization program. Include the people with responsibility for production, facilities, maintenance, quality control, and waste treatment and disposal on the team. It may be that a single person, such as the plant manager, has all of these responsibilities at a small facility. However, even at a small facility, at least two people should be involved to ensure some variety of viewpoints and perspectives.

Some larger companies have developed a system in which assessment teams periodically visit different facilities within the company. The benefits result through sharing the ideas and experiences with other divisions. Similar results can be achieved with periodic in-house seminars, workshops, and meetings. A large chemical manufacturer held a corporatewide symposium in 1986 dealing specifically with waste minimization. The company has also developed other programs to increase companywide awareness of waste minimization, including an internally published newsletter and videotape.

Careful planning and organization is necessary to bring about a successful waste minimization assessment program. To start the program and maintain momentum and control, it is necessary to obtain management commitment. The program should set general goals by which to measure its effectiveness. Selecting a good assessment program staff is critical to the ultimate success of the program. Since the assessment program is a project organization within the company, a task force provides an effective way of carrying out the program.

The "audit" or assessment step serves to identify the best options for minimizing waste through a thorough understanding of the waste-generating processes, wastestreams, and operating procedures. Therefore, the assessment teams' first major tasks are to identify, characterize, and track the facility's wastestreams and to compile information about the facility's waste-generating area, processes, and operations. These tasks can be carried out simultaneously.

5.1.4 Setting goals

The first priority of the waste minimization program task force is to establish goals that are consistent with the policy adopted by management. Waste minimization goals can be qualitative, such as "a significant reduction of toxic substance emissions into the environment." However, it is better to establish measurable, quantifiable goals, since qualitative goals can be interpreted ambiguously. Quantifiable goals establish a clear guide as to the degree of success expected of the program. For example, a major chemical company has adopted a corporatewide goal of 5 per cent waste reduction per year. In addition, each facility within the company has set its own waste minimization goals. As another example, as part of its general policy on hazardous waste, a large defense contractor has established an ambitious corporatewide goal of zero discharge of hazardous wastes from its facilities by the end of 1988. Each division within the corporation is given the responsibility and freedom to develop its own program (with intermediate goals) to meet this overall goal. This has resulted in an extensive investigation of procedures and technologies to accomplish source reduction, recycling, and resource recovery. It is important for a company's overall waste minimization goals to be incorporated into the appropriate individual departmental goals.

The goals of a company's waste minimization program should be reviewed periodically. As the focus of the program becomes more defined, the goals should be revised to reflect any changes. Waste minimization assessment is not intended to be a one-time project. Periodic reevaluation of goals is recommended due to the possibility of changes in available technology, raw material supplies, environmental regulations, economic climate, and so forth.

5.1.5 Overcoming barriers

The program task force should recognize potential barriers as it sets goals for waste minimization and then defines specific objectives that can be achieved. Although waste minimization projects can reduce operating costs and improve environmental compliance, they can also lead to conflicts between different groups, and jurisdictional conflicts can arise during the implementation of a waste minimization project.

In addition to jurisdictional conflicts related to these objectives, there are attitude-related barriers that can disrupt a waste minimization program. The commonly held attitude "if it ain't broke, don't fix it" stems from the desire to maintain the status quo and avoid the unknown. It also is based on the fear that a new waste minimization option may not work as advertised. Without the commitment to carefully conceive and implement an option, this attitude can become a self-fulfilling prophecy. Management must declare that "it *is* broke."

Another attitude-related barrier is the feeling that "it just won't work." This response is often given when a person does not fully understand the nature of the proposed option and its impact on operations. The danger here is that promising options may be dropped before they can be evaluated. One way to avoid this is to use idea-generating sessions (e.g., brainstorming). This encourages participants to propose a large number of options which are individually evaluated on their merits.

Another often-encountered barrier is the fear that a proposed waste minimization option will diminish product quality. This is particularly common in situations where unused feed materials are recovered from the waste and then recycled back to the process. The deterioration of product quality can be a valid concern if unacceptable concentrations of waste materials build up in the system. The best way to allay this concern is to set up a small-scale demonstration in the facility or to observe the particular option in operation at another facility.

Here are some examples of barriers to waste minimization, organized according to area or department:

Production

- A new operating procedure will reduce waste but may also produce a bottleneck that decreases the overall production rate.
- Production will be stopped while the new process equipment is installed.
- A new piece of equipment has not been demonstrated in a similar service. It may not work here.

Facilities/maintenance

- Adequate space is not available for installation of new equipment.

- Adequate utilities are not available for the new equipment.
- Engineering or construction personnel will not be available in time to meet project schedule.
- Extensive maintenance may be required.

Quality control

- More intensive quality control may be needed.
- More rework may be required.

Client relations and marketing

- Changes in product characteristics may affect customer acceptance.
- Changes in process may require recertification by customer or government agency.

Inventory

- A program to reduce inventory (to avoid material deterioration and reprocessing) may lead to stockouts during high product demand.

Finance

- There is not enough money to fund the project.

Purchasing

- Existing stocks (or binding contracts) will delay the replacement of a hazardous material with a nonhazardous substitute.

Environmental

- Accepting another plant's waste as a feedstock may require a lengthy resolution of regulatory issues.

Waste treatment

- Use of a new nonhazardous raw material will adversely impact the existing wastewater treatment facility.

5.1.6 Monitoring and Evaluation

Progress monitoring and evaluation is an integral component in every comprehensive improvement program. In the case of waste minimization, systematic monitoring should provide the following figures:

- Generation rates (or quantities) of emissions of concern, for every generating unit
- Costs of waste management, including treatment, handling, and disposal
- Production rates for every generating unit

The generation rates information can be obtained from manifests, emission inventories, and release estimates (e.g., toxic chemical release inventories required by Section 313, Title III of Superfund Amendments and Reauthorization Act of 1986). The costs of waste management are typically inclusive of handling, transportation, taxes, fees, treatment, and disposal. Production rates are necessary in order to normalize waste generation rate with respect to production rate so as to allow tracking of waste minimization process. Additionally, the monitoring effort should encompass all waste minimization projects and initiatives and document their costs and performance. Only then does it become possible to evaluate the effectiveness of the overall program and to rectify deficiencies.

5.2   Assessment Phase

The purpose of the assessment phase (see Fig. 5.5) is to develop a com-

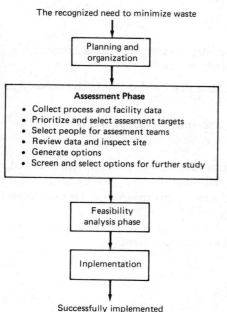

Figure 5.5   Assessment phase.

prehensive set of waste minimization options and to identify those that deserve additional, more detailed analysis. In order to develop these waste minimization options, a detailed understanding of the plant's wastes and operations is required. The assessment should begin by examining and analyzing information about the processes, operations, and waste management practices at the facility. Each of the assessment steps is discussed in this chapter.

5.2.1 Collecting and compiling information

The questions that this information-gathering effort will attempt to answer include the following:

- What wastestreams are generated from the plant? How much waste is generated?
- Which processes or operations do these streams come from?
- Which wastes are classified as hazardous? What makes them hazardous? Which wastes are not hazardous?
- What are the input materials used that generate the wastestreams of a particular process or plant area?
- How much of a particular input material enters each wastestream?
- How much of an input material can be accounted for through fugitive losses?
- How efficient is the process?
- Are unnecessary wastes generated by mixing otherwise recyclable hazardous wastes with other process wastes?
- What types of housekeeping practices are used to limit the quantity of wastes generated?
- What types of process controls are used to improve process efficiency?

Information that can be useful in conducting the assessment is listed below. Reviewing this information will provide important background for understanding the plant's production and maintenance processes and will allow priorities to be determined.

Facility information for waste minimization assessments

Design information

- Process flow diagrams
- Material application diagrams

- Piping and instrumentation diagrams
- Equipment lists, specifications, drawings, data sheets, operating and maintenance manuals
- Plot plans, arrangement drawings, and work flow diagrams

Environmental information

- Hazardous waste manifests
- Emission inventories and waste assays
- Biennial hazardous waste reports
- Environmental (compliance) audit reports
- Permits and permit applications
- Spill/release prevention and countermeasure plans

Raw material/production information

- Composition sheets
- Material safety data sheets
- Batch sheets
- Product and raw material inventory records
- Production schedules
- Operator data logs

Economic information

- Waste treatment and disposal costs
- Product, utility, and raw material costs
- Operating and maintenance labor costs

Other information

- Company environmental policy statements
- Company and department standard operating procedures
- Organization charts

Wastestream records.   One of the first tasks of a waste minimization assessment is to identify and characterize the facility wastestreams. Information about wastestreams can come from a variety of sources. Some information on waste quantities is readily available from the completed hazardous waste manifests, which include the description and quantity of hazardous waste shipped to a TSDF. The total amount of hazardous waste shipped during a one-year period, for example, is a convenient means of measuring waste generation and waste reduction efforts. However, manifests often lack such information as chemical analysis of the waste, specific source of the waste, and the time period during which the waste was generated. Also, manifests do not cover wastewater effluents, air emissions, or nonhazardous solid wastes.

Other sources of information on wastestreams include biennial reports and National Pollutant Discharge Elimination System (NPDES) monitoring reports. These NPDES monitoring reports will include the volume and constituents of wastewaters that are discharged. Additionally, toxic substance release inventories prepared under the right-to-know provisions of Section 313, Title III of Superfund Amendment and Reauthorization Act of 1986, may provide valuable information on emissions into all environmental media (land, water, and air). Also, analytical test data available from previous waste evaluations and routine sampling programs can be helpful if the focus of the assessment is a particular chemical within a wastestream.

Flow diagrams and material balances.   Flow diagrams provide basic means for identifying and organizing information that is useful for the assessment. Flow diagrams should be prepared to identify important process steps and to identify sources where wastes are generated, thus providing the foundation upon which material balances are built.

Material balances are important for waste minimization projects, since they allow for quantifying losses or emissions that were previously unaccounted for. Also, material balances assist in developing the following information:

- Baseline for tracking progress of the waste minimization efforts
- Data to estimate the size and cost of additional equipment, piping, instrumentation, and other materials
- Data to evaluate economic performance

In its simplest form, the material balance is represented by the mass conservation principle:

$$\text{Mass in} = \text{mass out} + \text{mass accumulated}$$

The material balance should be made individually for all components that enter and leave the process. When chemical reactions take place in a system, there is an advantage to doing "elemental balances" for specific chemical elements in a system.

Material balances can assist in determining concentrations of waste constituents where analytical test data is limited. They are particularly useful where there are points in the production process where it is difficult (due to inaccessibility) or uneconomical to collect analytical data. A material balance can help determine the magnitude of fugitive losses. For example, the evaporation of solvent from a parts cleaning tank can be estimated as the difference between solvent put into the tank and solvent removed from the tank.

To characterize wastestreams by material balance can require considerable effort, but it results in a more complete picture of the waste situation. This helps to establish the focus of the waste minimization activities and provides a baseline for measuring performance. By definition, the material balance includes both materials entering and leaving a process. Sources of material balance information include

- Samples, analyses, and flow measurements of feedstocks, products, and wastestreams
- Raw material purchase records
- Material inventories
- Emission inventories
- Equipment cleaning and validation procedures
- Batch makeup records
- Product specifications
- Design material balances
- Production records
- Operating logs
- Standard operating procedures and operating manuals
- Waste manifests

Material balances are easier, more meaningful, and more accurate when they are done for individual units, operations, or processes. For this reason, it is important to define the material balance envelope properly. The envelope should be drawn around the specific area of concern, rather than around a larger group of areas or the entire facility. An overall material balance for a facility can be constructed from individual unit material balances. This effort will highlight interrelationships between units and will help to point out areas for waste minimization by way of cooperation between different operating units or departments.

Pitfalls in preparing material balances. There are several factors that must be considered when preparing material balances, in order to avoid errors that could significantly overstate or understate wastestreams. The precision of analytical data and flow measurement of these quantities may be greater in magnitude than the actual wastestream itself. In this case, a reliable estimate of the wastestream cannot be obtained by subtracting the quantity of hazardous material in the product from that in the feed.

Time span is important when constructing a material balance. Material balances constructed over a shorter time span require more accurate and more frequent stream monitoring in order to close the balance. Material balances performed over the duration of a complete production run are typically the easiest to construct and are reasonably accurate. Time duration also affects the use of raw material purchasing records and on-site inventories for calculating input material quantities. The quantities of materials purchased during a specific time period should be monitored, since purchased materials can accumulate in warehouses or stockyards.

Developing material balances around complex processes can be a complicated undertaking, especially if recycle streams are present. Such tasks are usually performed by chemical engineers, often with the assistance of process simulation software.

Tracking wastes.   Changes in waste generation cannot be meaningful unless the information is collected both before and after a waste minimization option is implemented. Fortunately, it is easier to do material balances the second time, and it gets even easier as more are done because of the "learning curve" effect. In some larger companies, computerized data base systems have been used to track wastes. Figure 5.6 shows a worksheet used to record pertinent wastestream characteristics.

5.2.2   Prioritizing wastestreams and operations for assessment

Ideally, all wastestreams and plant operations should be assessed. However, prioritizing the wastestreams and operations for assessment is necessary when available funds and/or personnel are limited. The waste minimization assessments should concentrate on the most important waste problems first, and then move on to the lower priority problems as the time, personnel, and budget permit. Setting priorities as to which wastestreams or facility areas to assess requires a great deal of care and attention, since this step focuses the remainder of the assessment activity. Typical considerations for prioritizing wastestreams include

- Compliance with current and future regulations

Firm	Waste minimization assessment	Prepared by
Site	Proc. unit oper.	Checked by
Date	Proj. no.	Sheet 1 of 1 Page

Worksheet		Waste stream summary

Attribute		Stream no.		Stream no.		Stream no.	
Waste ID/name:							
Source/origin							
Component or property of concern							
Annual generation rate (units)							
Overall							
Component(s) of concern							
Cost of disposal							
Unit cost ($ per:)							
Overall (per year)							
Method of management[2]							

Priority rating criteria[3]	Relative wt. (W)	Rating (R)	R × W	Rating (R)	R × W	Rating (R)	R × W
Regulatory compliance							
Treatment/disposal cost							
Potential liability							
Waste quantity generated							
Waste hazard							
Safety hazard							
Minimization potential							
Potential to remove bottleneck							
Potential by-product recovery							
Sum of priority rating scores		Σ(R × W)		Σ(R × W)		ΣR × W)	
Priority rank							

Notes:
1. Stream numbers, if applicable, should correspond to those used on process flow diagrams.
2. For example, sanitary landfill, hazardous waste landfill, on-site recycle, incineration, combustion with heat recovery, distillation, dewatering, etc.
3. Rate each stream in each category on a scale from 0 (none) to 10 (high).

Figure 5.6 Wastestream summary.

- Costs of waste management (treatment and disposal)
- Potential environmental and safety liability
- Quantity of waste
- Hazardous properties of the waste (including toxicity, flammability, corrosivity, and reactivity)

- Other safety hazards to employees
- Potential for removing bottlenecks in production or waste treatment
- Potential recovery of valuable by-products
- Available budget for the waste minimization assessment program and projects

Small businesses and large businesses with only a few waste-generating operations should assess the entire facility. It is also beneficial to look at an entire facility when there are a large number of similar or identical operations. Similarly, the implementation of good operating practices that involve procedural or organizational measures, such as soliciting employee suggestions, awareness-building programs, better inventory and maintenance procedures, and internal cost accounting changes, should be implemented on a facilitywide basis. Since many of these options do not require large capital expenditures, they should be implemented as soon as is practical.

5.2.3 Site inspection

With a specific area or wastestream selected, and with the assessment team in place, the assessment continues with a visit to the site. In the case where the entire assessment team is employed at the plant being assessed, members of the team should familiarize themselves with the site as much as possible. Although the collected information is critical to gaining an understanding of the processes involved, seeing the site is important in order to witness the actual operation. Very frequently, a process unit is operated differently from the method originally described in the operating manual. Also, modifications may have been made to the equipment that were not recorded in the flow diagrams, equipment lists, or other documentation.

When people from outside of the plant participate in the assessment, it is recommended that a formal site inspection take place. Even when the team is made up entirely of plant employees, a site inspection by all team members is helpful after the site information has been collected and reviewed. The inspection helps to resolve questions or conflicting data uncovered during the review. The site inspection also provides additional information to supplement that obtained earlier.

When the assessment team includes members employed outside of the plant, the team should prepare a list of needed information on an inspection agenda. The list can be presented in the form of a checklist detailing objectives, questions and issues to be resolved, and further information requirements. The agenda and information list are given to the appropriate plant personnel in the areas to be assessed early enough before the visit to allow them to assemble the information in advance. Of course, it may be that the assessment team members themselves are in the best position to collect and compile much of the data. If the agenda and needs list are carefully thought out, important points are less likely to be overlooked during the inspection. Some guidelines for the inspection are as follows:

- Prepare an agenda in advance that covers all points that still require clarification. Provide staff contacts in the area being assessed with the agenda several days before the inspection.

- Schedule the inspection to coincide with the particular operation that is of interest (e.g., makeup chemical addition, batch sampling, batch dumping, start-up, shutdown, etc.).

- Monitor the operation at different times during the shift and, if needed, during all three shifts, especially when waste generation is highly dependent on human involvement (e.g., in painting or parts cleaning operations).

- Interview the operators, shift supervisors, and forepersons in the assessed area. Do not hesitate to question more than one person if an answer to a concern or inquiry is not forthcoming. Assess the operators' and supervisors' awareness of the waste generation aspects of the operation. Note their familiarity (or lack thereof) with the impacts their operation may have on other operations.

- Photograph the area of interest, if warranted. Photographs are valuable in the absence of plant layout drawings. Many details can be captured in photographs that otherwise could be forgotten or inaccurately recalled at a later date.

- Observe the housekeeping aspects of the operation. Check for signs of spills or leaks. Visit the maintenance shop and ask about any problems in keeping the equipment leak-free. Assess the overall cleanliness of the site. Pay attention to odors and fumes.

- Assess the organizational structure and level of coordination of environmental activities between various departments.

- Assess administrative controls, such as cost accounting procedures, material purchasing procedures, and waste collection procedures.

In performing the site inspection the assessment team should follow the process from the point where raw materials enter the area to the point where the products and the wastes leave the area. The team should identify the suspected sources of waste. This may include the production process, maintenance operations, storage areas for raw materials, finished product, and work in process. Recognize that the plant's waste treatment area itself may also offer opportunities to minimize waste. This inspection often results in forming preliminary conclusions about the causes of waste generation. Full confirmation of these conclusions may require additional data collection, analysis, and site visits.

5.2.4 Generating options

Once the origins and causes of waste generation are understood, the assessment process enters the creative phase. The objective of this step is to generate a comprehensive set of waste minimization options for further consideration. Following the collection of data and site inspections, the members of the team will have begun to identify possible ways to minimize waste in the assessed area. Identifying potential options relies both on the expertise and creativity of the team members. Much of the requisite knowledge may come from their education and on-the-job experience; however, the use of technical literature, contacts, and other sources is always helpful.

The process for identifying options should follow a hierarchy in which source reduction options are explored first, followed by recycling options. This hierarchy of effort stems from the environmental desirability of source reduction as the preferred means of minimizing waste. Treatment options should be considered only after acceptable waste minimization techniques have been identified.

The ultimate aim of source reduction is to avoid the generation of wastes altogether, thereby eliminating the question of what to do with the wastes after they are born. Source reduction techniques are characterized as good operating practices, technology or process changes, input material changes, and product changes. The aim of recycling is to put process residuals to a beneficial use. Recycling techniques are characterized as use/reuse techniques and resource recovery techniques. Recycling can be performed off site or on site.

The process by which waste minimization options are identified should occur in an environment that encourages creativity and independent thinking by the members of the assessment team. While the individual team members will suggest many potential options on their own, the process can be enhanced by using some of the common group-decision techniques. These techniques allow the assessment team to identify options that the individual members might not have come up with on their own. Brainstorming sessions with the team members are an effective way of developing waste minimization options. Figs. 5.7 and 5.8 are examples of worksheets used to develop an options list.

Firm _____	**Waste minimization assessment**	Prepared by _____
Site _____	Proc. unit oper. _____	Checked by _____
Date _____	Proj. no. _____	Sheet 1 of 1 Page ___ of ___

	Option generation

Meeting format (e.g., brainstorming, nominal group technique) _____
Meeting coordinator _____
Meeting participants _____
_____
_____
_____

List suggested options	Rationale/remarks on option

Figure 5.7 Option generation.

5.2.5 Screening options

Many waste minimization options will be identified in a successful assessment. At this point, it is necessary to identify those options that offer real potential to minimize waste and reduce costs. Since detailed evaluation of technical and economic feasibility is usually costly,

Firm _____	**Waste minimization assessment**	Prepared by _____
Site _____	Proc. unit oper. _____	Checked by _____
Date _____	Proj. no. _____	Sheet _1_ of _1_ Page ___ of ___

Worksheet	Option description

Option name: _____

Briefly describe the option. _____

_____

_____

_____

_____

_____

_____

_____

Wastestream(s) affected: _____

Input material(s) affected: _____

Product(s) affected: _____

Indicate type: ☐ Source reduction

_____ Equipment-related change

_____ Personnel/procedure-related change

_____ Materials-related change

☐ Recycling/reuse

_____ On site   _____ Material reused for original purpose

_____ Off site   _____ Material used for a lower-quality purpose

_____ Material sold

_____ Material burned for heat recovery

Originally proposed by: _____ Date: _____

Reviewed by: _____ Date: _____

Approved for study? _____ yes _____ no, by: _____

Reason for acceptance or rejection _____

Figure 5.8 Option description.

the proposed options should be screened to identify those that deserve further evaluation. The screening procedure serves to eliminate suggested options that appear to be marginal, impractical, or inferior, without a detailed and more costly feasibility study.

The screening procedures can range from an informal review and a decision made by the program manager or a vote of the team members, to quantitative decision-making tools. The informal evaluation is an unstructured procedure by which the assessment team or waste minimization program task force selects the options that appear to be the best. This method is especially useful in small facilities, with small management groups, or in situations where only a few options have been generated. This method consists of a discussion and examination of each option.

The weighted sum method (see Fig. 5.9) is a means of quantifying the important factors that affect waste management at a particular facility and the way each option will perform with respect to these factors. This method is recommended when there are a large number of options to consider.

No matter what method is used, the screening procedure should consider the following questions:

- What is the main benefit gained by implementing this option? (e.g., economics, compliance, liability, workplace safety, etc.)
- Does the necessary technology exist to develop the option? What development effort is required?
- How much does it cost? Is it cost-effective?
- Can the option be implemented within a reasonable amount of time without disrupting production?
- Does the option have a good track record? If not, is there convincing evidence that the option will work as required?

The results of the screening activity are used to promote the successful options for technical and economic feasibility analyses. The number of options chosen for the feasibility analyses depends on the time, budget, and resources available for such a study.

Some options (such as procedural changes) may involve no capital costs and can be implemented quickly with little or no further evaluation. The screening procedure should account for ease of implementation of an option. If such an option is clearly desirable and indicates a potential cost savings, it should be promoted for further study or outright implementation.

5.3 Feasibility Analysis

The final product of the assessment phase is a list of waste minimization options postulated for the assessed area. The assessment will have screened out the impractical or unattractive options. The next step is to determine if the remaining options are technically and economically feasible. Figure 5.10 depicts the feasibility analysis phase.

Firm _____
Site _____
Date _____

Waste minimization assessment
Proc. unit oper. _____
Proj. no. _____

Prepared by _____
Checked by _____
Sheet __1__ of __1__ Page __1__ of __1__

Worksheet

Option evaluation by weighted sum method

Criteria	Weight (W)	Options rating (R)													
		#1 Option		#2 Option		#3 Option		#4 Option		#5 Option	
		R	R × W	R	R × W	R	R × W	R	R × W	R	R × W
Reduction in waste's hazard											
Reduction of treatment/disposal costs											
Reduction of safety hazards											
Reduction of input material costs											
Extent of current use in industry											
Effect on product quality (no effect = 10)											
Low capital cost											
Low O & M cost											
Short implementation period											
Ease of implementation											

Final evaluation	Sum of weighted ratings $\Sigma (W \times R)$
	Option ranking
	Feasibility analysis scheduled for (date)

Figure 5.9 Weighted sum method.

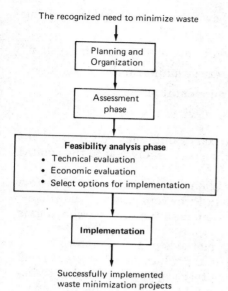

Figure 5.10 Feasibility analysis phase.

5.3.1 Technical evaluation

The technical evaluation determines whether a proposed waste minimization option will work in a specific application. The assessment team should use a "fast track" approach in evaluating procedural changes that do not involve a significant capital expenditure. Process testing of materials can be done relatively quickly, if the options do not involve major equipment installation or modifications.

For equipment-related options or process changes, visits to see existing installations can be arranged through equipment vendors and industry contacts. The operator's comments are especially important and should be compared with the vendor's claims. Bench-scale or pilot-scale demonstration is often necessary. Often it is possible to obtain scale-up data using a rental test unit for bench-scale or pilot-scale experiments. Some vendors will install equipment on a trial basis, with acceptance and payment after a prescribed time, if the user is satisfied. Typical technical evaluation criteria include the following:

- Is the system safe for workers?
- Will product quality be maintained?
- Is the new equipment, material, or procedure compatible with production operating procedures, work flow, and production rates?
- Is additional labor required?

- Are utilities available? Or must they be installed, thereby raising capital costs?
- How long will production be stopped in order to install the system?
- Is special expertise required to operate or maintain the new system?
- Does the system create other environmental problems?
- Does the vendor provide acceptable service?

Although an inability to meet these constraints may not present insurmountable problems, correcting them will likely add to the capital and/or operating costs.

All affected groups in the facility should contribute to and review the results of the technical evaluation. Prior consultation and review with the affected groups (e.g., production, maintenance, purchasing) is needed to ensure the viability and acceptance of an option. If the option calls for a change in production methods or input materials, the project's effects on the quality of the final product must be determined. If, after the technical evaluation, the project appears infeasible or impractical, it should be dropped. Figure 5.11 is an example worksheet showing important items to consider when evaluating the technical feasibility of a waste minimization option.

5.3.2   Economic evaluation

As in any project, the cost elements of a waste minimization project can be broken down into capital costs and operating costs. The economic evaluation is then carried out using standard measures of profitability, such as payback period, return on investment, and net present value. Each organization has its own economic criteria for selecting projects for implementation. The reader is referred to standard textbooks on engineering economics (such as Peters and Timmerhaus, 1980) for details of calculations and profitability analysis.

For smaller facilities with only a few processes, the entire waste minimization assessment procedure will tend to be much less formal. In this situation, several obvious waste minimization options, such as installation of flow controls and good operating practices, may be implemented with little or no economic evaluation. In these instances, no complicated analyses are necessary to demonstrate the advantages of adopting the selected waste minimization options. A proper perspective must be maintained between the magnitude of savings that a potential option may offer, and the amount of personnel, time, and effort required to do the technical and economic feasibility analyses.

Firm _____	**Waste minimization assessment**	Prepared by _____
Site _____	Proc. unit oper. _____	Checked by _____
Date _____	Proj. no. _____	Sheet _1_ of _6_ Page ___ of ___

| Worksheet | **Technical feasibility** |

WM option description _____

1. Nature of WM option
   - ☐ Equipment-related
   - ☐ Personnel/procedure-related
   - ☐ Material-related

2. If the option appears technically feasible, state your rationale for this. _____
   _____
   _____

   Is further analysis required? ☐ Yes ☐ No. If yes, continue with this
   worksheet. If not, skip to worksheet 15.

3. Equipment-related option

	Yes	No.	
Equipment available commercially?	☐	☐	_____
Demonstrated commercially?	☐	☐	_____
in similar application?	☐	☐	_____
Successfully?	☐	☐	_____

   Describe closest industrial analog. _____
   _____
   _____

   Describe status of development. _____
   _____
   _____

Prospective vendor	Working installation(s)	Contact person(s)	Date contracted*

*Also attach filled out phone conversation notes, installation visit report, etc.

Figure 5.11   Technical feasibility.

5.3.3 Adjustments for risks and liability

As mentioned earlier, waste minimization projects may reduce the magnitude of environmental and safety risks for a company. Although these risks can be identified, it is difficult to predict the likelihood that problems will occur, the nature of the problems, and their resulting magnitude. One way of accounting for the reduction of these risks is to ease the financial performance requirements of the project. For

Firm _____	**Waste minimization assessment**	Prepared by _____
Site _____	Proc. unit oper. _____	Checked by _____
Date _____	Proj. no. _____	Sheet _2_ of _6_ Page ___ of ___

Worksheet	**Technical feasibility**
	(continued)

WM option description _____

3. Equipment-related option (continued)

Performance information required (describe parameters): _____

_____

_____

_____

_____

Scaleup information required (describe): _____

_____

_____

_____

_____

Testing required: ☐ yes   ☐ no

   Scale: ☐ bench ☐ pilot ☐ _____

   Test unit avaliable? ☐ yes ☐ no _____

   Test parameters (list) _____

_____

_____

_____

Number of test runs: _____

Amount of material(s) required: _____

Testing to be conducted: ☐ In-plant

                             ☐ _____

Facility/product constraints:

   Space requirements _____

   Possible locations within facility _____

Figure 5.11 *Continued.*

example, the acceptable payback may be lengthened from four to five years, or the required internal rate of return may be lowered from 15 percent to 12 percent. Such adjustments reflect recognition of elements that affect the risk of exposure of the company but cannot be included directly in the analyses. These adjustments are judgmental and necessarily reflect the individual viewpoints of the people evaluating the project for capital funding. Therefore, it is important that

Firm _____	**Waste minimization assessment**	Prepared by _____
Site _____	Proc. unit oper. _____	Checked by _____
Date _____	Proj. no. _____	Sheet __4__ of __6__ Page ___ of ___

Worksheet	**Technical feasibility**
	(continued)

WM option description _____

3. Equipment-related option (continued)

Will modifications to work flow or production procedures be required? Explain. _____

_____

_____

_____

_____

Operator and maintenance training requirements

Number of people to be trained _____ ☐ On site

_____ ☐ Off site

Duration of training

Describe catalyst, chemicals, replacement parts, or other supplies required.

Item	Rate or frequency of replacement	Supplier, address

Does the option meet government and company safety and health requirements?

☐ Yes ☐ No Explain _____

_____

_____

How is service handled (maintenance and technical assistance)? Explain _____

_____

_____

_____

What warranties are offered? _____

_____

_____

Figure 5.11 *Continued.*

Firm _____	Waste minimization assessment	Prepared by _____
Site _____	Proc. unit oper. _____	Checked by _____
Date _____	Proj. no. _____	Sheet _3_ of _6_ Page ___ of ___

Worksheet	Technical feasibility
	(continued)

WM option description _____

2. Equipment-related option (continued)

    Utility requirements:

Electric power	Volts (AC or DC) _____	kW _____
Process water	Flow _____	Pressure _____
	Quality (tap,/demin, etc.) _____	
Cooling water	Flow _____	Pressure _____
	Temp. in _____	Temp. out _____
Coolant heat transfer fluid _____		
	Temp. in _____	Temp. out _____
	Duty _____	
Steam	Pressure _____	Temp. _____
	Duty _____	Flow _____
Fuel	Type _____	Flow _____
		Duty _____
Plant air _____	Flow _____	
Inert gas _____	Flow _____	

Estimated delivery time (after award of contract) _____

Estimated installation time _____

Installation dates _____

Estimated production down time _____

Will production be otherwise affected? Explain the effect and impact on production. _____

_____

_____

_____

Will product quality be effected? Explain the effect on quality. _____

_____

_____

_____

Figure 5.11 *Continued.*

the financial analysts and the decision makers in the company be aware of the risk reduction and other benefits of the waste minimization options. As a policy to encourage waste minimization, some companies have set lower hurdle rates for waste minimization projects. The reader is referred to the next chapter for more in-depth discussion

Firm _____	**Waste minimization assessment**	Prepared by _____
Site _____	Proc. unit oper. _____	Checked by _____
Date _____	Proj. no. _____	Sheet _5_ of _6_ Page ___ of ___

Worksheet	**Technical feasibility**
	(continued)

WM description _____

3. Equipment-related option (continued)

   Describe any additional storage or material handling requirements. _____
   _____
   _____
   _____

   Describe any additional laboratory or analytical requirements. _____
   _____
   _____
   _____

4. Personnel/procedure-related changes
   Affected departments/areas _____
   _____
   _____

   Training requirements _____
   _____
   _____
   _____

   Operating instruction changes. Describe responsible departments. _____
   _____
   _____
   _____

5. Materials-related changes (*Note*: If substantial changes in equipment are required, then handle the
   option as an equipment-related one.)   Yes   No

   Has the new material been demonstrated commercially?   ☐   ☐
   In a similar application?   ☐   ☐
   Successfully?   ☐   ☐

   Describe closest application. _____
   _____
   _____

Figure 5.11 *Continued.*

Firm_____	**Waste minimization assessment**	Prepared by _____
Site _____	Proc. unit oper. _____	Checked by _____
Date _____	Proj. no. _____	Sheet _6_ of _6_ Page ___ of ___

Worksheet	**Technical feasibility**
	(continued)

WM option description _____

4.   Materials-related changes (continued)

Affected departments/areas _____

_____

_____

_____

Will production be affected? Explain the effect and impact on production.

_____

_____

_____

Will product quality be affected? Explain the effect and the impact on product quality.

_____

_____

_____

_____

Will additional storage, handling, or other ancillary equipment be required? Explain.

_____

_____

_____

Describe any training or procedure changes that are required.

_____

_____

_____

Describe any material testing program that will be required.

_____

_____

_____

Figure 5.11 *Continued.*

of how intangible components, such as liability, enter the economic feasibility analyses.

5.4 Implementation

Figure 5.12 shows the elements in the implementation phase. The waste minimization assessment report provides the basis for obtaining company funding of waste minimization projects. Because projects are not always sold on their technical merits alone, a clear description of both tangible and intangible benefits can help edge a proposed project past competing projects for funding.

The champions of the waste minimization assessment program should be flexible enough to develop alternatives or modifications. They should also be committed to the point of doing background and support work, and they should anticipate any potential problems in implementing the options. Above all, they should keep in mind that an idea will not sell if the sponsors of the idea are not sold on it themselves.

5.4.1 Funding

Waste reduction projects generally involve improvements in process efficiency and/or reductions in operating costs of waste management. However, an organization's capital resources may be prioritized to-

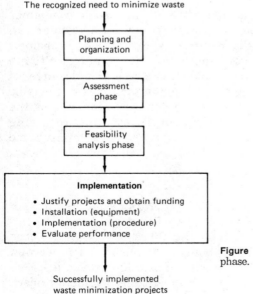

Figure 5.12 Implementation phase.

ward enhancing future revenues (for example, moving into new lines of business, expanding plant capacity, or acquiring other companies), rather than toward cutting current costs. If this is the case, then a sound waste reduction project could be postponed until the next capital budgeting period. It is then up to the project sponsor to ensure that the project is reconsidered at that time.

Knowing the level within the organization that has approval authority for capital projects will help in enlisting the appropriate support for waste minimization projects. In large corporations, smaller projects are typically approved at the plant manager level, medium-sized projects at the divisional vice president level, and larger projects at the executive committee level.

An evaluation team made up of financial and technical personnel can ensure that a sponsor's enthusiasm is balanced with objectivity. It can also serve to quell opposing "can't be done" or "if it ain't broke, don't fix it" attitudes that might be encountered within the organization. The team should review the project in the context of

- Past experience in this area of operation
- What the market and the competition are doing
- How the implementation program fits into the company's overall business strategy
- Advantages of the proposal in relation to competing requests for capital funding

Even when a project promises a high rate of return, some companies will have difficulty raising funds internally for capital investment. In this case, the company generally has two major sources to consider: private sector financing and government-assisted funding. Private sector financing includes bank loans and other conventional sources of financing. Government financing is available in some cases. It may be worthwhile to contact your state's Department of Commerce or the federal Small Business Administration for information regarding loans for pollution control or hazardous waste disposal projects. Some states can provide technical and financial assistance.

5.4.2 Installation

Waste minimization options that involve operational, procedural, or materials changes (without additions or modifications to equipment) should be implemented as soon as the potential cost savings have been determined. For projects involving equipment modifications or new equipment, the installation of a waste minimization project is essentially no different from any other physical plant improvement project.

The phases of the project include planning, design, procurement, and construction.

5.4.3 Demonstration and follow-up

After the waste minimization option has been implemented, it remains to be seen how effective the option actually turns out to be. Options that don't measure up to their original performance expectations may require rework from vendors prior to installation of the equipment. The documentation provided through a follow-up evaluation represents an important source of information for future uses of the option in other facilities. Figure 5.13 is a form for evaluating the performance of an implemented waste minimization option and documenting the experience gained in implementing options at subsequent facilities.

5.4.4 Waste reduction measurement

One measure of effectiveness for a waste minimization project is the project's effect on the organization's cash flow. The project should pay for itself through reduced waste management costs and reduced raw materials costs. However, it is also important to measure the actual reduction of waste accomplished by the waste minimization project. The easiest way to measure waste reduction is by recording the quantities of waste generated before and after a waste minimization project has been implemented. The difference, divided by the original waste generation rate, represents the percentage reduction in waste quantity. However, this simple measurement ignores other factors that also affect the quantity of waste generated.

In general, waste generation is directly dependent on the production rate. Therefore, the ratio of waste generation rate to production rate is a convenient way of measuring waste reduction. Expressing waste reduction in terms of the ratio of waste to production rates is not free of problems, however. One of these problems is the danger of using the ratio of infrequent large quantities to the production rate. This problem is illustrated by a situation where a plant undergoes a major overhaul involving equipment cleaning, paint stripping, and repainting. Such overhauls are fairly infrequent and are typically performed every three to five years. The decision to include this intermittent stream in the calculation of the waste reduction index, based on the ratio of waste rate to production rate, would lead to an increase in this index. This decision cannot be justified, however, since the infrequent generation of painting wastes is not a function of production rate. In a situation like this, the waste reduction progress should be measured

Firm	Waste minimization assessment	Prepared by
Site	Proc. unit oper.	Checked by
Date	Proj. no.	Sheet _1_ of _1_ Page ___ of ___

Worksheet	Option performance

WM option description _____

☐ Baseline   ☐ Projected   ☐ Actual
(without option)

(a)   Period duration _____ From _____ To _____
(b)   Production per period _____ Units(_____)
(c)   Input materials consumption per period

Material	Pounds	Pounds/unit product

(d)   Waste generation per period

Wastestream	Pounds	Pounds/unit product

(e)   Substance(s) of concern: Generation rate per period

Wastestream	Substance	Pounds	Pounds/unit product

Figure 5.13 Option performance.

in terms of the ratio of waste quantity or materials use to the square footage of the area painted. In general, a distinction should be made between production-related wastes, maintenance-related wastes, and cleanup wastes. Also, a few wastestreams may be inversely proportional to production rate. For example, a waste resulting from outdated input materials is likely to increase if the production rate decreases. This is because the age-dated materials in inventory are more likely to expire when their use in production decreases. For these reasons, care must be taken when expressing the extent of waste reduction. This requires that the origins of waste generation be well understood.

In measuring waste reduction, the total quantity of an individual wastestream should be measured, as well as the individual waste components or characteristics. Many companies have reported substantial reduction in the quantities of waste disposed. Often, much of the reduction can be traced to good housekeeping and steps taken to concentrate a dilute aqueous waste. Although concentration, as such, does not fall within the definition of waste minimization, there are practical benefits that result from concentrating wastewater streams, including decreased disposal costs. Concentration may render a wastestream easier to recycle, and concentration is also desirable if a facility's current wastewater treatment system is overloaded.

Obtaining good-quality data for wastestream quantities, flows, and composition can be costly and time-consuming. For this reason, it may be practical, in some instances, to express waste reduction indirectly in terms of the ratio of input materials consumption to production rate. These data are easier to obtain, although the measure is not direct.

Measuring waste minimization by using a ratio of waste quantity to material throughput of product output is generally more meaningful for specific units or operations, rather than for an entire facility. Therefore, it is important to preserve the focus of the waste minimization project when measuring and reporting progress. For those operations not involving chemical reactions, it may be helpful to measure waste minimization progress by using the ratio of input material quantity to material throughput or production rate.

The planning and design team for a new product, production process, or operation should address waste generation aspects early on. The assessment procedure in this manual can be modified to provide a waste minimization review of a product or process in the planning or design phase. The earlier such an assessment is performed, the less likely it is that the project will require expensive changes. All new projects should be reviewed by the waste minimization program task force.

The waste minimization program is a continuing, rather than a one-time effort. Once the highest priority wastestreams and facility areas have been assessed and those projects have been implemented, the assessment program should look to areas and wastestreams with lower priorities. The ultimate goal of the waste minimization program should be to reduce the generation of waste to the maximum extent achievable. Companies that have eliminated the generation of hazardous waste should continue to look at reducing industrial wastewater discharges, air emissions, and solid wastes.

The frequency with which assessments are done will depend on the program's budget, the company's budgeting cycle (annual cycle in most companies), and the presence of special circumstances, such as:

- A change in raw material or product requirements
- Higher waste management costs
- New regulations
- New technology
- A major event with undesirable environmental consequences (such as a major spill)

Aside from the special circumstances, a new series of assessments should be conducted each fiscal year.

To be truly effective, a culture of waste minimization must be developed in the organization. This means that waste minimization must be an integral and vital part of the company's operations.

Tables 5.1 through 5.7 contain useful information that will aid in generating and evaluating options for potential waste minimization opportunities. In order to develop a list of waste minimization options for a facility, it is necessary to understand the sources, causes, and controlling factors that influence waste generation, as presented in Tables 5.1 and 5.2.

TABLE 5.1   Typical Wastes from Plant Operations

Plant function	Location/operation	Potential waste material
Material receiving	Loading docks, incoming pipelines, receiving areas	Packaging materials, off-specification materials, damaged containers, inadvertant spills, transfer hose emptying
Raw material and product storage	Tanks, warehouses, drum storage yards, bins, storerooms	Tank bottoms; off-specification and excess materials; spill residues; leaking pumps, valves, tanks, and pipes; damaged containers; empty containers
Production	Melting, curing, baking, distilling, washing, coating, formulating, reaction	Washwater; rinse water; solvents; still bottoms; off-specification products; catalysts; empty containers; sweepings; ductwork cleanout; additives; oil; filters; spill residue; excess materials; process solution dumps; leaking pipes, valves, hoses, tanks, and process equipment
Support services	Laboratories	Reagents, off-specification chemicals, samples, empty sample and chemical containers
	Maintenance shops	Solvents, cleaning agents, degreasing sludges, sand-blasting waste, caustics, scrap metal, oils, greases
	Garages	Oils, filters, solvents, acids, caustics, cleaning bath sludges, batteries
	Powerhouses/boilers	Fly ash, slag, tube cleanout material, chemical additives, oil, empty containers, boiler blow-down, water-treating chemical wastes
	Cooling towers	Chemical additives, empty containers, cooling tower bottom sediment, cooling tower blowdown, fan lube oils

SOURCE: Adapted from Gary Hunt and Roger Schecter, "Minimization of Hazardous Waste Generation," in *Standard Handbook of Hazardous Waste Management*, Harry Freeman (ed.), McGraw-Hill, New York.

TABLE 5.2   Causes and Controlling Factors in Waste Generation

Waste origin	Typical causes	Operational factors	Design factors
Chemical reaction	Incomplete conversion By-product formation Catalyst deactivation (by poisoning or sintering)	Inadequate temperature control Inadequate mixing Poor feed flow control Poor feed purity control	Proper reactor design Proper catalyst selection Choice of process Choice of reaction conditions
Contact between aqueous and organic phases	Condensate from steam jet ejectors Presence of water as a reaction by-product Use of water for product rinse Equipment cleaning Spill cleanup	Indiscriminate use of water for cleaning or washing	Vacuum pumps instead of steam jet ejectors Choice of process Use of reboilers instead of steam stripping
Process equipment cleaning	Presence of cling Deposit formation Use of filter aids Use of chemical cleaners	Drainage prior to cleaning Production scheduling to reduce cleaning frequency	Design reactors or tanks wiper blades Reduce cling Equipment dedication
Heat exchanger cleaning	Presence of cling (process side) or scale (cooling water side) Deposit formation Use of chemical cleaners	Inadequate cooling water treatment Excessive cooling water temperature	Design for lower film temperature and high turbulence Controls to prevent cooling water from overheating
Metal parts cleaning	Disposal of spent solvents, spent cleaning solution, or cleaning sludge	Indiscriminate use of solvent or water	Choice between cold dip tank or vapor degreasing Choice between solvent or aqueous cleaning solution

Waste origin	Typical causes	Operational factors	Design factors
Metal surface treating	Drag-out Disposal of spent treating solution	Poor rack maintenance Excessive rinsing with water Fast removal of workpiece	Countercurrent rinsing Fog rinsing Drag-out collection tanks or trays
Disposal of unusable raw materials or off-specification products	Obsolete raw materials Off-specification products caused by contamination, improper reactant controls, inadequate precleaning of equipment or workpiece, temperature or pressure excursions	Poor operator training or supervision Inadequate quality control Inadequate production planning and inventory control of feedstocks	Use of automation Maximize dedication of equipment to a single function
Cleanup of spills and leaks	Manual material transfer and handling operations Leaking pump seals Leaking flange gaskets	Inadequate maintenance Poor operator training Lack of attention by operator Excessive use of water in cleaning	Choice of gasketing materials Choice of seals Use of welded or seal-welded construction

SOURCE: Jacobs Engineering Group

TABLE 5.3 Waste Minimization Options for Equipment Cleaning Operations

Waste	Source/origin	Waste reduction measures	Remarks	References
Spent solvent- or inorganic-based cleaning solutions	Tank cleaning operations	Maximize dedication of process equipment.		
		Use squeegees to recover cling of product prior to rinsing.		
		Avoid unnecessary cleaning.		
		Use closed storage and transfer systems.	Scaling and drying up can be prevented.	
		Provide sufficient drain time for liquids.	Minimizes leftover material.	
		Line the equipment to prevent cling.	Reduces cling.	18
		"Pig" the process lines.		
		Use high-pressure spray nozzles.		19
		Use countercurrent rinsing.		
		Use clean-in-place systems.	Minimizes solvent consumption.	
		Clean equipment immediately after use.	Prevents hardening of scale that requires more severe cleaning.	

Waste	Process	Waste reduction measures		Ref.
Wastewater sludges, spent acidic solutions	Heat exchanger cleaning	Reuse cleanup solvent.		
		Rework cleanup solvent into useful products.		
		Segregate wastes by solvent type.		
		Standardize solvent usage.		
		Reclaim solvent by distillation.		
		Schedule production to lower cleaning frequency.		
		Use bypass control or pumped recycle to maintain turbulence during turndown.	On-site or off-site recycling.	
		Use smooth heat-exchange surfaces.	Electroplated or Teflon tubes.	20
		Use on-stream cleaning techniques.	"Superscrubber," for example.	21
		Use hydroblasting over chemical cleaning where possible.		

TABLE 5.4 Waste Minimization Options for Coating Operations

Waste	Source/origin	Waste reduction measures	Remarks	References
Coating overspray	Coating material that fails to reach the object being coated	Maintain 50% overlap between spray pattern.	The coated object does not look streaked, and wastage of coating material is avoided. If the spray gun is arched 45°, the overspray can be as high as 65%.	1,2
		Maintain 6″–8″ distance between spray gun and workpiece.		
		Maintain a gun speed of about 250 ft/min.		
		Hold gun perpendicular to the surface.		
		Trigger gun at the beginning and end of each pass.		
		Assure proper training of operators.		2
		Use robots for spraying.		2
		Avoid excessive air pressure for coating atomization.	By air pressure adjustment, overspray can be reduced to 40%.	2
		Recycle overspray.		3
		Use electrostatic spray systems.	Overspray can be reduced by 40%.	4

Waste source	Source	Prevention option	Benefit	Ref.
Stripping wastes	Coating removal from parts before applying a new coat	Use air-assisted airless spray guns in place of air-spray guns.	Increases transfer efficiency.	4
		Avoid adding excess thinner.	Reduces stripping wastes due to rework.	5
		Use abrasive media stripping.	Solvent usage is eliminated.	6
		Use bead-blasting for paint stripping.	Solvent usage is eliminated.	7
		Use cryogenic stripping.	Solvent usage is eliminated.	8
		Use caustic stripping solutions.	Solvent usage is eliminated.	1
		Clean coating equipment after each use.		
Solvent emissions	Evaporative losses from process equipment and coated parts	Keep solvent soak tanks away from heat sources.		9
		Use high-solids formulations.	Lower usage of solvents.	
		Use powder coatings.	Avoids solvent usage.	10,11
		Use water-based formulations.	Avoids solvent usage.	4,12

TABLE 5.4 Waste Minimization Options for Coating Operations (Continued)

Waste	Source/origin	Waste reduction measures	Remarks	References
Equipment cleanup wastes	Process equipment cleaning with solvents	Use light-to-dark batch sequencing.		13
		Produce large batches of similarly coated objects instead of small batches of differently coated items.		
		Isolate solvent-based paint spray booths from water-based paint spray booths.		20
		Reuse cleaning solutions and solvent.		
		Standardize solvent usage.		
Overall		Reexamine the need for coating, as well as available alternatives.		

TABLE 5.5 Waste Minimization through Good Operating Practices

Good operating practice	Program ingredients	Remarks	References
Waste minimization assessments	Form a team of qualified individuals Establish practical short-term and long-term goals Allocate resources and budget for the program Establish assessment targets Identify and select options to minimize waste Periodically monitor the program's effectiveness	These programs are conducted to reduce waste in a facility.	22
Environmental audits/reviews	Assemble pertinent documents Conduct environmental process reviews Carry out a site inspection Report on and follow up on the findings	These audits are conducted to monitor compliance with regulations.	23,24
Loss prevention programs	Establish Spill Prevention, Control, and Countermeasures (SPCC) plans Conduct hazard assessment in the design and operating phases	SPCC plans are required by law for oil storage facilities.	3,25,26
Waste segregation	Prevent mixing of hazardous wastes with nonhazardous wastes Isolate hazardous wastes by contaminant Isolate liquid wastes from solid wastes	These measures can result in lower waste haulage volumes and easier disposal of the hazardous wastes.	4
Preventive maintenance programs	Use equipment data cards on equipment location, characteristics, and maintenance Maintain a master preventive maintenance (PM) schedule Deferred PM reports on equipment Maintain equipment history cards Maintain equipment breakdown reports Keep vendor maintenance manuals handy Maintain a manual or computerized repair history file	These programs are conducted to cut production costs and decrease equipment downtime, in addition to preventing waste releases due to equipment failure.	27,28,29

TABLE 5.5 Waste Minimization through Good Operating Practices (Continued)

Good operating practice	Program ingredients	Remarks	References

Training/awareness-building programs	Provide training for:
	  Safe operation of the equipment
	  Proper materials handling
	  Economic and environmental ramifications of
	    hazardous waste generation and disposal
	  Detecting releases of hazardous materials
	  Emergency procedures
	  Use of safety gear
		These programs are conducted to reduce occupational health and safety hazards, in addition to reducing waste generation due to operator or procedural errors.	2

Effective supervision	Closer supervision may improve production efficiency and reduce inadvertent waste generation. Management by objectives (MBO), with goals for waste reduction.	Increased opportunity for early detection of mistakes. Better coordination among the various parts of an overall operation.	
Employee participation	Quality circles (free forums between employees and supervisors) can identify ways to reduce waste. Solicit employee suggestions for waste reduction ideas.	Employees who intimately understand the operations can identify ways to reduce waste.	
Production scheduling/planning	Maximize batch size. Dedicate equipment to a single product. Alter batch sequencing to minimize cleaning frequency (light-to-dark batch sequence, for example). Schedule production to minimizing cleaning frequency.	Altering production schedule can have a major impact on waste minimization.	
Cost accounting/allocation	Cost accounting done for all wastestreams leaving the facilities. Allocate waste treatment and disposal costs to the operations that generate the waste.	Allocating costs to the waste-producing operations will give them an incentive to cut their wastes.	

TABLE 5.6 Waste Minimization Options in Materials Handling, Storage, and Transfer

Waste/source	Waste reduction measures	Remarks	References
Material/waste tracking and inventory control	Avoid overpurchasing. Accept raw material only after inspection. Ensure that inventory quantity does not go to waste. Ensure that no containers stay in inventory longer than a specified period. Review material procurement specifications. Return expired material to supplier. Validate shelf-life expiration dates. Test outdated material for effectiveness. Eliminate shelf-life requirements for stable compounds. Conduct frequent inventory checks. Use computer-assisted plant inventory system. Conduct periodic materials tracking. Label all containers properly. Set up manned stations for dispensing chemicals and collecting wastes.	These procedures are employed to find areas where the waste minimization efforts are to be concentrated.	30,31
Loss prevention programs	Use properly designed tanks and vessels for their intended purposes only. Install overflow alarms for all tanks and vessels. Maintain physical integrity of all tanks and vessels. Set up written procedures for all loading/unloading and transfer operations. Install secondary containment areas. Forbid operators to bypass interlocks and alarms or to alter setpoints significantly without authorization. Isolate equipment or process lines that leak or are not in service. Use seal-less pumps. Use bellows-seal valves. Document all spillage. Perform overall material balances and estimate the quantity and dollar value of all losses. Use floating-roof tanks for VOC control. Use conservation vents on fixed roof tanks. Use vapor recovery systems.		

TABLE 5.6 Waste Minimization Options In Materials Handling, Storage, and Transfer(*Continued*)

Waste/source	Waste reduction measures	Remarks	References
Spills and leaks	Store containers in such a way as to allow for visual inspection for corrosion and leaks.		
	Stack containers so as to minimize the chance of tipping, puncturing, or breaking.		
	Prevent concrete "sweating" by raising the drum off storage areas.		
	Maintain Material Safety Data Sheets (MSDSs) to correctly handle spill situations.		
	Provide adequate lighting in storage areas.		
	Maintain a clean, even surface in transportation areas.		
	Keep aisles clear of obstruction.		
	Maintain distance between incompatible chemicals.		
	Maintain distance between different types of chemicals to prevent cross-contamination.		
	Avoid stacking containers against process equipment.		
	Follow manufacturers' suggestions on the storage and handling of all raw materials.		
	Insulate and inspect electric circuitry for corrosion and potential sparking.		
Cling	Use large containers instead of small containers whenever possible.		
	Use containers with height-to-diameter ratio equal to one to minimize wetted area.		
	Empty drums and containers thoroughly before cleaning or disposal.		

TABLE 5.7 Waste Minimization Options for Parts Cleaning Operations

Waste	Source/origin	Waste reduction measures	Remarks	References
Spent solvent	Contaminated solvent from parts cleaning operations	Use water-soluble cutting fluids instead of oil-based fluids.	This could eliminate the need for solvent cleaning.	
		Use peel coatings in place of protective oils.		
		Use aqueous cleaners.		
		Use aqueous paint-stripping solutions.	8	
		Use cryogenic stripping.	7	
		Use bead blasting for paint stripping.	6	
		Use multistage countercurrent cleaning.		
		Prevent cross-contamination.		
		Prevent drag-in from other processes.		
		Remove sludge from tank promptly.		
		Reduce the number of different solvents used.	A single, larger waste that is more amenable to recycling.	
Air emissions	Solvent loss from degreasers and cold tanks	Use roll-type covers, not hinged covers.	24 to 50% reduction in emissions.	15
		Increase freeboard height.	39% reduction in solvent emissions.	15
		Install freeboard chillers.		15
		Use silhouette entry covers.		
		Use proper equipment layout.		
		Avoid rapid insertion and removal of items.	The speed that items are put into the tank should be less than 11 ft/min.	16

TABLE 5.7 Waste Minimization Options for Parts Cleaning Operations (Continued)

Waste	Source/origin	Waste reduction measures	Remarks	References
		Avoid inserting oversized objects into the tank.	Cross-sectional area of the item should be less that 50% of tank area to reduce piston effect.	17
		Allow for proper drainage before removing an item.		
		Avoid water contamination of solvent in degreasers.		
Rinse water	Water rinse to remove solvent carried out with the parts leaving the cleaning tank	Reduce solvent drag-out by proper design and operation of rack system.	The drag-out can be 0.4 gal/1000 sq ft, versus 24 gal/1000 sq ft for poorly drained parts.	15
		Install air jets to blow parts dry. Use fog nozzles on rinse tanks.		
		Ensure proper design and operation of barrel system.		15
		Use countercurrent rinse tanks.		15
		Use water sprays on rinse tanks.	More efficient rinsing is achieved.	15

124

REFERENCES

1. J. Kohl, J. Pearson, and P. Wright, *Managing and Recycling Solvents in the Furniture Industry*, North Carolina State University, Raleigh, N.C., 1986.
2. D. Lenckus, "Increasing Productivity," *Finishing Wood and Wood Products Magazine*. vol. 87, no. 4, May 1982, pp. 44–66.
3. M. E. Campbell and W. M. Glenn, *Profit from Pollution Prevention*, The Pollution Probe Foundation, Toronto, Canada, 1982.
4. J. Kohl, P. Moses, and B. Triplett, *Managing and Recycling Solvents: North Carolina Practices, Facilities, and Regulations*. North Carolina State University, Raleigh, N.C., 1984.
5. J. J. Durney, "How to Improve Your Paint Stripping," *Product Finishing*, December 1982, pp. 52–53.
6. T. E. Higgins, *Industrial Process Modifications to Reduce Generation of Hazardous Waste at DOD Facilities: Phase I Report*. CH2M Hill, Washington, D.C., 1985.
7. "Cryogenic Paint Stripping," *Product Finish*, December 1982.
8. W. M. Mallarnee, "Paint and Varnish Removers," *Kirk-Othmer Encyclopedia of Chemical Technology*, 3d ed., vol. 16, 1981, pp. 762–767.
9. J. Sandberg, *Final Report on the Internship served at Gage Tool Company*, Minnesota Technical Assistance Program, Minnesota Waste Management Board, Minnesota, 1985.
10. Powder Coatings Institute, Information brochure, Washington, D. C., 1983.
11. G. E. Cole, "VOC Emission Reduction and Other Benefits Achieved by Major Powder Coating Operations," paper no. 84-38. 1 presented at the Air Pollution Control Association, June 25, 1984.
12. California State Department of Health Services, *Alternative Technology for Recycling and Treatment of Hazardous Waste*, 3d Biennial Report, Sacramento, 1986.
13. California State Department of Health Services, *Guide to Solvent Waste Reduction Alternatives*, October 1986, pp. 4–25 to 4–49.
14. R. E. Kenson, "Recovery and Reuse of Solvents from VOC Air Emissions," *Environmental Progress*, August 1985, pp. 161–165.
15. L. J. Durney (ed.), *Electroplating Engineering Handbook*, 4th ed., Van Nostrand Reinhold, New York, 1984.
16. American Society of Testing Materials, *Handbook of Vapor Degreasing*, Special Technical Publication 310-A., ASTM, Philadelphia, April 1976.
17. C. Smith, "Troubleshooting Vapor Degreasers," *Product Finish*, November 1981.
18. C. M. Loucks, "Boosting Capacities with Chemicals," *Chemical Engineering Deskbook Issue*, vol. 80, no. 5, 1973, pp. 79–84.
19. 3M Corporation, *Ideas—A Compendium of 3M Success Stories*, St. Paul, Minn.
20. C. H. Fromm, S. Budaraju, and S. A. Cordery, "Minimization of Process Equipment Cleaning Waste," Conference proceedings of HAZTECH International, Denver, August 13–15, 1986, pp. 291–307.
21. Versar, Inc. and Jacobs Engineering Group, *Waste Minimization: Issues and Options*, vol. II, U. S. Environmental Protection Agency, Washington, D. C., October 1986.
22. C. H. Fromm and M. S. Callahan, "Waste Reduction Audit Procedure," Conference proceedings of the Hazardous Materials Control Research Institute, Atlanta, 1986, pp. 427–435.
23. North Carolina Pollution Prevention Pays Program, *Environmental Auditing*, North Carolina Department of Environmental Health, 1985.
24. R. A. Baumer, "Making environmental audits," *Chemical Engineering*, vol. 89, no. 22, November 1, 1982, p. 101.
25. T. A. Kletz, "Minimize Your Product Spillage," *Hydrocarbon Processing*, vol. 61, no. 3, 1982, p. 207.
26. D. Sarokin, "Reducing Hazardous Wastes at the Source: Case Studies of Organic Chemical Plants in New Jersey," paper presented at Source Reduction of Hazardous Waste Conference, Rutgers University, August 22, 1985.

27. J. B. Singh and R. M. Allen, "Establishing a Preventive Maintenance Program," *Plant Engineering*, February 27, 1986, p. 46.
28. D. Rimberg, "Minimizing Maintenance Makes Money," *Pollution Engineering*, vol. 12, no. 3, December 1983, p. 46.
29. N. H. Parker, "Corrective Maintenance and Performance Optimization," *Chemical Engineering*, vol. 91, no. 7, April 16, 1984, p. 93.
30. E. Geltenan, "Keeping Chemical Records on Track," *Chemical Business*, vol. 6, no. 11, 1984, p. 47.
31. W. E. Hickman and W. D. Moore, "Managing the Maintenance Dollar," *Chemical Engineering*, vol. 93, no. 7, April 24, 1986, p. 68.
32. "The EPA Manual for Wash Minimization Assessments," U.S. Environmental Protection Agency, Washington, D.C.

The Economics of
Waste Minimization

Ronald T. McHugh

Office of Policy, Planning, and Evaluation
U.S. Environmental Protection Agency

Introduction

In Chapter 5, which discussed waste minimization assessments,* one of the concepts presented was a screening analysis of both technical and economic variables during the assessment step to identify the most promising options. The assessment team would screen options against such variables as the ease of implementing a particular option, the cost and liability reduction potential, and the track record of the option. Having identified the leading options, the assessment

---

*I refer to pollution prevention and waste minimization almost interchangeably in the text. *Pollution prevention* (through source reduction and recycling) is the more accurate term, since it conveys that we are interested in reducing waste releases to all media—not just to the wastestreams. Simply put, we want to reduce the loadings which the environment actually "sees."

team would then perform a more detailed analysis of technical and economic feasibility. This chapter presents a general approach to doing that more detailied economic analysis.

As was made clear in the chapter about waste minimization assessment, the economics of waste minimization have frequently been the driving force in process engineering evolving from the early time-and-motion studies, through material shortages in World War II, to the energy crisis of the 1970s, and finally to today's emphasis on compliance and materials cost savings tied to the direct and indirect costs of environmental requirements. While the various rationales (productive efficiency, energy conservation, etc.) for prevention projects have changed over time, the need to base a prevention project on sound economics has not. It is important to understand that the economics of prevention will continue to change over time as competition, materials and energy costs, disposal costs, regulatory costs, and liability costs continue to evolve.

In the context of waste minimization, an analogy to a trip to the dentist is somewhat appropriate: Your company may want to incur preventive costs today to reduce current production, energy, and materials costs along with current and future regulatory, insurance, and/or liability costs. This chapter gives only the fundamentals of how to carry out such an analysis of private costs (waste-related costs incurred or avoided by a specific company, i.e., not including the social or environmental costs of environmental releases), since a complete presentation of all potential costs avoided through a prevention program would both require process-specific data and be well beyond the scope of this chapter.

6.2   What Is the Economics of Waste Minimization?

Economic analysis of waste minimization is a matter of determining how to calculate the costs and benefits of a prevention option; how to compare costs and benefits to current practice, when the timing of both costs and benefits may vary over a number of years; how to calculate a statistic such as net present value (NPV) to rank a project's financial viability; and how to understand and use what this NPV number represents to help make decisions. This chapter recommends (1) regarding the entire plant as a "system" for prevention and minimization opportunities, (2) implementing those minimization options that appear to be worthwhile, and (3) reevaluating the viability of all options after implementing the first.

The format for this chapter is to take you briefly through the why and the when questions and then on to a more lengthy discussion of the how-to questions about economic analysis, together with an example involving a hypothetical electroplating operation. Much of the basic framework for this chapter and the example is drawn from the U.S. Environmental Protection Agency's (EPA's) *Pollution Prevention Benefits Manual** which describes all of these calculation procedures much more completely.

6.3 Why Analyze the Economics of Prevention?

A waste minimization or pollution prevention program has to compete for company resources (dollars and people) against other worthwhile projects from all around the company, some of which may be "mandatory" investments such as installing a waste treatment unit. To enable a prevention project to compete with other investments, an assessment team will need to demonstrate not only that it has thoroughly analyzed both the technical and the economic feasibility, but also that the numbers used in analysis are as "robust" as they can be.

If your firm is just beginning a prevention program, the opportunities should be many, with substantial monetary benefits from each. This chapter should help identify the obvious *and* the less obvious benefits. As you begin to exhaust the "low-hanging fruit" options, a point of diminishing returns will eventually be reached. At this point, options become fewer and more costly to implement, while at the same time incremental benefits, i.e., cost savings from prevention, will fall. For most firms, this point of diminishing returns is, however, a long way off and will continue to evolve.

It is important to recognize that an analysis at a point in time captures only a snapshot of costs and benefits. Materials costs, liability costs, and so on will continue to evolve, so that a "correct" decision in 1990 may not be a "correct" decision in 1992.

6.4 What Is the General Analysis Procedure?

Generally, the amount of detail that goes into an analysis reflects how far the company or plant has gone into pollution prevention in the past. If the firm is just starting, many projects can be shown to be cost-beneficial by measuring only directly avoided treatment and disposal costs and improvements in product yield (i.e., units of output per

---

Pollution Prevention Benefits Manual—Phase II Draft, prepared by ICF, Inc., Fairfax, Va. Available, USEPA, 401 M Street SW, Washington, D.C. 20460.

pound of input), or what I term *Tier 0* of analysis.* As the company goes further in pursuing options (after having demonstrated some successes on which to build), the assessment team will probably need to expand the types of cost savings or project benefits being considered.† That is, the person doing the analysis should do only as much detailed calculation as is necessary to support the project, but should recognize that the assessment team and company managers may want to consider broadening the list beyond Tier 0 to Tiers 1, 2, or perhaps even 3, which are described briefly in this section and more fully below. Again, a Tier 0 analysis will frequently be sufficient if you are just beginning to implement your prevention program. Tier 0 analysis can be performed by looking at such variables as input loss during storage or production and improved materials handling practice that can lead to greater production per unit of time and/or unit of input. Such cost savings can also result if materials are being recycled and thus producing positive revenue rather than requiring costly special handling and storage as "wastes."

Figure 6.1 illustrates the procedure. At each incremental "tier" of the analysis you compare the estimated costs of the prevention project (capital and O&M) to a broadening scope of avoided capital and O&M costs (benefits) resulting from the prevention project over current operations. Those costs and benefits are then translated to a financial measure, e.g., NPV. Once you have determined if the project is cost-beneficial at Tier 0 you can determine whether you should broaden the analysis to Tiers 1, 2, and 3. The additional tiers (levels) progressively differ as to (1) how likely it is that a specific plant will encounter particular costs (e.g., Resource Conservation and Recovery Act [RCRA] corrective action) and need to be able to limit them, (2) the accuracy with which a company can estimate the costs avoided in that tier, and (3) the possible impact on the company (what is at risk) if the effect is encountered. Thus, as the analysis proceeds further towards Tier 3, both certainty of cost occurrence and precision in estimating avoidable costs fall, although dollar risk rises. Tier 1 includes Tier 0 benefits plus avoided regulatory costs, Tier 2 adds avoided liability costs to the Tier 1 figure, and Tier 3 adds avoided intangible costs to the Tier 2 figure.

Tier 1 looks at everyday regulatory costs which the firm can limit

---

*As a brief aside, experts on prevention agree that you should index your waste reduction achievements against specific plant output, e.g., "pounds of waste per automobile manufactured." A good waste minimization program may actually experience an increase in emissions or wastes if production increases dramatically.

†All project costs and cost savings (benefits) can be categorized as either capital or operating and maintenance (O&M). In the discussion of the tables and various tiers of analysis which follow, this categorization will be maintained.

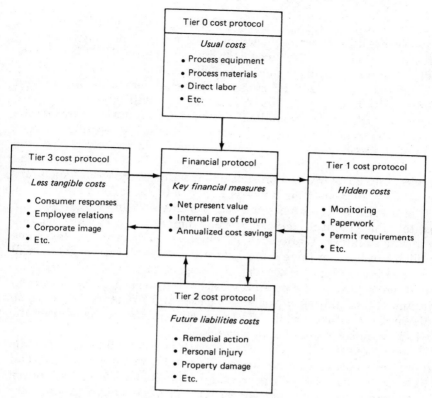

Figure 6.1 Estimation of waste minimization benefits.

through minimization and prevention. Such costs as record keeping, reporting, testing and monitoring, and manifesting are obvious elements that a company can limit, for example by moving a hazardous waste treatment unit inside a production process and recycling the material before it becomes a waste or perhaps by limiting waste residuals to such an extent (< 220 lb/month) so as to become a nonregulated very small quantity generator. Less obvious examples of Tier 1 actions might be to avoid some charges currently incurred by the company for releasing excess metals to the publicly owned treatment works (POTW) or to switch to an aqueous nonhalogenated solvent for parts cleaning, thereby avoiding costs by complying with the U.S. Occupational Safety and Health Administration (OSHA) requirements on worker exposure and also avoiding RCRA waste management and liability costs.

Tier 2 analysis projects benefit from avoiding fines or failures that may be catastrophic—such as the failure of a liner or other potential liabilities—i.e. occurrence and detection of a failure somewhere in the environmental control system. The landfill, with the attendant need to perform a RCRA corrective action or more mundane—e.g., violating a National Pollution Discharge Elimination System (NPDES) permit. Tier 2 analysis involves placing a dollar value on such potentially avoided liability costs. It is obvious that if members of the assessment team choose to employ a Tier 2 analysis they will have to deal with somewhat subjective probabilities of various events actually occurring. At the same time, to ignore system failures in analyzing the benefits of a prevention option seems foolhardy. This is especially true since, for example, EPA projects most landfill liners to fail after approximately 20 years. The present dollar value of the cost of such failures—even 20 years from now—is likely to be quite large.

Tier 3 involves yet a more subjective framework in evaluating even less tangible potential impacts, such as company or corporate risk and financial standing. While such parameters as favorable/unfavorable publicity, corporate image and goodwill, product acceptance, overall corporate or director risk, and the impact on the firm's financial standing because of the possibility of having to report significant environmental liabilities on a U.S. Securities and Exchange Commission 10K form are all intangible effects, such variables can have dramatic and sometimes irreversible impacts.

Some analysts will be comfortable with only Tier 0 or Tier 1 while others can embrace Tier 2 more readily and be prepared to utilize Tier 3 as a "swing variable" if the costs and benefits are close. The bottom line? There is no "right" answer for how far to go, but the assessment team should weigh the costs of collecting more information and analysis versus the value that it brings to saving the company money.

6.5   What Information Do We Need to Start?

During the assessment preparation step, and again in fleshing out the options for further study, the assessment team should have developed a great deal of information on solid and hazardous wastestreams and other broadly defined "emissions" categories, together with relevant process and plant operating data for current practice and at least one alternative. Table 6.1 is intended to provide a screening-level checklist for the team which can subsequently be used during the feasibility assessment phase to record the economic information. Obviously, not all baseline practices or prevention alternatives will have an impact on every cost category. Note that only variables for Tier 0 and Tier 1 are presented, with the less obvious Tier 1 regulatory costs described as "hidden."

As an example of this approach, a firm interested in substituting a material which produces a nonhazardous wastestream for a currently

TABLE 6.1   Economic Information Checklist—Tier 0

	Baseline	Alternative	Years of cash flow
Capital Costs			
Purchased process equipment			
Materials			
Utility connections			
Additional equipment			
Site preparation			
Installation			
Engineering and procurement			
Start-up cost			
One-time training costs			
Permitting costs			
Initial charge of catalysts and chemicals			
Fixed capital investment			
Working capital			
Total capital investment			
Salvage value			
Net capital investment for Tier 0			
Operating Cost/Revenue Item			
Change in direct treatment, disposal cost			
Change in raw materials cost			
Change in materials handling and transportation costs			
Change in utilities cost			
Change in catalysts and chemicals			
Change in O&M labor costs			
Change in O&M supplies			
Change in insurance costs			
Change in other operating costs			
Incremental revenues from increased (decreased) production			
Incremental revenues from marketable by-products			
Net operating cost savings for Tier 0			

SOURCE: Taken from EPA's *Opportunity Assessment Manual.*

utilized raw material which yields a RCRA hazardous waste may find that it can lower both direct *and* hidden (regulatory) capital costs through a prevention project. The current wastestream may necessitate installation of a new mandatory treatment unit in order to meet land disposal restrictions. Since this cost is mandatory (i.e., it could be regarded as Tier 0), the firm could compare this imminent capital cost decision with the capital equipment required to utilize a nonhazardous chemical. (As a practical matter, there is no "bright line" between Tier 0 and Tier 1, but which tier to focus on is a matter of choice for the assessment team which relates to the team's awareness of imminent changes in regulatory requirements.) Depending upon the other elements of the waste management system, the company may also find it can avoid both reporting and record keeping, as well as certain less obvious Tier 1 costs, such as the need to retrofit the liner system in a surface impoundment or to comply with air emissions from various units no longer receiving hazardous wastes. In performing a Tier 1 analysis you can work through the regulatory checklist appendix and consult Appendix B of the *Pollution Prevention Benefits Manual* for a brief description of certain regulatory requirements and their applicabilities.

6.6 What to Do?

After compiling whatever information is relevant on the checklist, the assessment team can calculate annualized costs and cash flows for the current practice and for the alternative on a sheet like Worksheet 0 (see Fig. 6.2).* (One worksheet is for the current practice, and one is for the alternative). These values are then combined and summarized for each succeeding tier on a form such as Worksheet V (see Fig. 6.3), which presents the net annualized savings *between* the current practice and the alternative. You should not at this point become overly concerned with understanding the technical meaning of annualized costs, discount rates, or net present values, since a later section of this chapter presents a commonsense explanation of these terms.

While the calculation of annualized costs and present values is somewhat tedious, it is not difficult, and it is necessary to be able to compare streams of costs and benefits that extend into the future. Thus, each entry in the annualized cost (5%, 10%, 15%) columns of Worksheet 0 represents a discounted average of a stream of costs or benefits from now until the end of the project period (frequently 20 years).

6.7 An Example

This section abstracts from and expands upon the hypothetical firm example in Appendix D in EPA's *Pollution Prevention Benefits Manual*. The example illustrates the economic analysis procedure for a fairly narrowly defined problem in a small-to-medium-sized single-plant firm. The firm is an electroplater of gold jewelry with 40 employees and gross revenues of $10 million. It is located in an industrial park on the outskirts of a large industrial city in which there are sev-

---

*Annualized cost is a concept that is best exemplified by a fixed monthly mortgage payment. The concept of annualized savings is best exemplified by the stream of future fixed payment you get on a bond for investing money today for a fixed period.

TIER 0 • *Usual Costs*   ☐ Current practice   ☐ Alternative practice

Item description	Cash flow information				Annualized costs*			
	Escalation rate (r_e, %)	First year of cashflow (t_1, years)	Last year of cashflow (t_2, years)	Cashflow estimate (C^*)	$r_d = 5\%$	$r_d = 10\%$	$r_d = 15\%$	$r_d =$
A. Depreciable capital expenditures								
A1 Equipment								
A2 Materials								
A3 Utility connections								
A4 Site preparation								
A5 Installation								
A6 Engineering and procurement								
B. Expenses								
B1 Start-up								
B2 Raw materials								
B3 Permitting								
B4 Initial catalysts								
B5 Working capital								
B6 Salvage value								
B7 Disposal								
B8 Raw materials								
B9 Utilities								
B10 Catalysts and chemicals								
B11 Labor								
B12 Supplies								
B13 Insurance								
B14 Other								
C. Operating revenues								
C1 Revenues								
C2 By-product revenues								

* *In thousands of year-0 dollars*

Figure 6.2   Worksheet O.

eral other jewelry electroplaters, many of whom have similar operations with respect to technologies and wastes.

Because of the concentration of jewelry electroplaters, the city POTW has fairly strict limits (e.g., 2.0 ppm copper) on discharge of metals in the wastewater to the POTW. Additionally, jewelry electroplaters have strict controls by the POTW due to the presence of cyanide-containing plating baths in which the metals to be plated are dipped. The bath itself is subject to continuous filtration, and the fil-

Financial Worksheet
(In thousands of year-0 dollars)   Tier ☐

Item description	Annualized savings			
Current less alternative	$r_d = 5\%$	$r_d = 10\%$	$r_d = 15\%$	$r_d =$
a. Depreciable capital expenditures				
b. Expenses				
c. Operating revenues				
d. Future liabilities				

e. Tax liabilities				

f. Total savings				

Aggregate savings

Tier 0	
Tier 1	
Tier 2	
Tier 3	

Figure 6.3   Worksheet V.

ters are changed twice per week. The filtrate (metals, slag, and dust) is less than 100 kg/month, so that this wastestream is not subject to RCRA Subtitle C requirements, i.e., this wastestream qualifies for very small quantity generator status. The spent plating bath is periodically neutralized, so only a small percentage of cyanide and metal is discharged in wastewater to the POTW.

However, prior to entering the plating bath, the parts to be plated are cleaned in a open-top vapor degreaser (OTVD) currently utilizing the chlorinated solvent 1,1,1-trichloroethane (TCA). The metals cleaning process generates approximately 2250 lb of a RCRA hazardous waste (FOO2) per month (2500 gal/year at 10.8 lb/gal). The wastestream consists of the TCA, with 5 ppm nickel and 10 percent suspended solids. The waste is collected below the process line and sent to an aboveground tank from which a commercial hazardous waste management firm collects it at least once every 90 days. Since the company stores the waste for 90 days or less, it does not require an RCRA storage permit. The commercial waste handler serves a number of electroplaters, recycling TCA through distillation, and manages the residual still bottoms by incineration and landfilling of the resulting ash in its RCRA Subtitle C permitted landfill.

The owner and manager of the hypothetical firm, Mr. Auric, has learned through a trade industry publication that there are some water-based cleaners that have performed well in jewelry plating operations. The same article mentioned that OSHA has proposed* to further limit worker exposure to several chlorinated solvents by lowering the permissible exposure limit (PEL) for several chlorinated solvents, although TCA is not one of the ones addressed. However, the trade association article indicates that beyond the new OSHA limits and existing RCRA treatment standards for solvents, EPA may also consider additional controls on several chlorinated solvents in cleaning operations. For some time, Mr. Auric has been concerned over the rising cost of managing chlorinated solvents, including TCA, through the commercial hazardous waste firm he contracts with.

Mr. Auric (a one-man assessment team) obtains screening information on the costs and performance characteristics of the aqueous cleaner with data showing that it performs equally as well as TCA does for his parts cleaning operation. The aqueous cleaner is less expensive than TCA, but its use would require a capital equipment purchase. Switching from TCA to an aqueous cleaner would also avoid RCRA requirements which Mr. Auric feels will ultimately become very costly and perhaps lead to liability.

Mr. Auric's first job is to gather the information he needs to do the feasibility (technical and economic) analysis. He assumes that the modifications would take place this year and that the existing cleaner would last for 20 years at normal maintenance levels ($1000 per year). The firm buys 24300 lb of TCA per year at $0.40 per lb ($9720 per year) and pays $5000 in disposal fees per year. Further, based on information from the manufacturer, Mr. Auric assumes a useful life of 20 years for the new aqueous cleaner.

The capital costs are $50,000 for the new cleaner and $5000 for installation. Annual operating costs are $1100 for materials, $500 for maintenance, and $3250 for utilities. He would not face any additional disposal costs since the water-based cleaner changes could be discharged to the POTW.

Mr. Auric's company faces a federal tax rate of 34 percent on profits with a 10 percent state corporate profits tax for a total of 44 percent. He discounts future costs back to the present value of those dollars at a 15 percent rate and assumes that inflation will average 4 percent over the period.† Further, his accountant informs him that the Tax

---

*53FR2197, June 7, 1988

†For most business planning, it is appropriate to use overall rates of inflation to ac-

TABLE 6.2   Benefits of Switching from Current Practice to PP Alternative

Level of analysis	Cost item	Net savings or benefits (in $ per year)*
TIER 0:	*Standard or Usual Capital and Operating & Maintenance Costs*	
	Equipment and installation	$ 8,790
	Raw materials	16,200
	Energy	9,910
	Disposal	5,140
	Labor	1,960
	Supplies	970
	Total savings through tier 0	$ 260
TIER 1:	*Hidden Regulatory Costs*	
	Reporting	$ 930
	Inspections	1,820
	Other	880
	Total savings through Tier 1	$ 103
TIER 2:	*Future Liabilities*	
	Treatment or storage in tank liability	$47,500
	Transportation liability	1,500
	Disposal in landfill liability	35,300
	Total savings through Tier 2	$84,403
TIER 3:	*Less Tangible Benefits*	
	Increase in operating revenues	$ 4,400
	Total savings through Tier 3	$88,803

*Before-tax estimates assuming a discount rate of 15 percent. Negative estimates represent a cost increase or net loss. Positive estimates represent a cost decrease or net benefit.

Reform Act of 1986 allows him to depreciate or write off new capital assets through an attractive "200 percent double declining balance" method over seven years. Table 6.2 summarizes the key variables.

Based on an analysis of project costs and benefits from Tier 0, Mr. Auric concludes that the project has a current value to him of – $260.* This means that he is acquiring debt for which he will be paying back $260 per year. Said another way, if someone gave him a grant of ex-

count for the most likely future operating and capital costs. Recognize that for RCRA-regulated hazardous wastes, treatment and disposal cost increases have and will continue to far exceed the general rate of inflation. For this reason, the worksheets allow you to employ different rates of inflation for any line item.

*Discounting the stream of future costs and benefits back to present values at 15 percent per year. At a lower discount rate, e.g., 5 or 10 percent, the project might exhibit positive returns.

Financial Worksheet
(In thousands of year–dollars)   Tier ☐ **0**

Item description	Annualized cash flow*			
Alternative less current	$r_d = 5\%$	$r_d = 10\%$	$r_d = 15\%$	$r_d = 11.14\%$
a. Depreciable capital expenditures	−4.41	−6.46	−8.79	−6.97
b. Expenses	8.92	8.66	8.53	8.62
c. Operating revenues	0	0	0	0
d. Future liabilities	—		—	

e. Tax liabilities	−2.28	−1.76	−1.33	−1.65

f. Net savings for tier	2.28	0.44	−1.58	0

Total savings | | | | | IRR
Through tier 0	2.28	0.44	−1.58	0	11.14%
Through tier 1					
Through tier 2					
Through tier 3					

* *Cashflow estimates for alternative less current, by definition, are the same as cash flow estimates for incremental analysis.*

Figure 6.4 Worksheet V (example problem).

actly $260 per year he would be indifferent to doing the project. However, Mr. Auric pursues the analysis into Tier 1 to explore his avoided regulatory costs if he implements the aqueous cleaning system and finds that the project is justified at Tier 1 (annualized savings of $103). The example will not be repeated from the benefits manual, but suffice it to say that he would save an additional $3630 in annual operating and maintenance costs (annualized to $103). As shown in Fig. 6.4, his aggregate savings from the project climb through Tier 2. Note that there is really no reason for him to go beyond Tier 1 in his analysis, since the project was justified at that level.

The key numbers presented in Table 6.2 are annualized cost and present values, discounted at 15 percent over the 20-year project period and calculated with the assumption that all costs go up at 4 percent per year, due to inflation. As noted above, a negative overall present value for the project at Tier 0 implies that you wouldn't do the project unless, for example, you got an annual grant of $260 from somewhere (such as the state or local government) to make it worthwhile. Once you expand the list of benefits to include regulatory costs which you can avoid by doing the project, the annualized savings and present value become positive. The NPV at Tier 1 implies that the project savings are equal to investing in a bond or any other investment producing an income stream of $103 per year over 20 years.

Waste Reduction and Business Planning

David Wigglesworth

Program Manager,
Waste Reduction Assistance Program (WRAP)
Alaska Health Project

Until recently, business owners were not overly concerned with the costs resulting from the generation of industrial waste. Today that situation has changed. Business owners and their employees now bear an enormous legal, financial, and social responsibility to manage safely the materials used in their business operations, in addition to any wastes produced. Developing sound business plans which properly manage and, where possible, eliminate these materials and wastes is critical in both starting a new venture and maintaining an existing one, especially for small to medium-sized companies with slim profit margins. Even seemingly small losses (wastes) can add up year after year, adversely affecting business profits, workplace health and safety, and the quality of our environment.

Examination of losses from the handling of industrial materials and wastes is often neglected when formulating and implementing a business plan. This chapter explores various methods for including waste reduction in a business plan. A well-developed business plan can help company managers reduce or eliminate all types of waste, including solid waste, such as paper, boxes, and other refuse; spills, vapors, and fumes that may cause pollution outdoors and in the workplace; wastewater discharges; energy waste; and hazardous wastes that are reactive, toxic, ignitable, and corrosive—to name just a few. Figure 7.1 in Sec. 7.3 provides interested readers with a "roadmap" for evaluating a business plan with a critical eye toward waste reduction opportunities.

7.1   What Is a Business Plan?

A business plan is a profile of all the components of your company. Making a business plan means putting onto paper a blueprint that turns your ideas into reality and helps maintain your operation for years to come. A business plan typically includes the following major elements:

1. *Marketing plan:*   This plan focuses on how to market your business and considers issues such as market supply and demand information and marketing strategies.

2. *Operations plan:*   This plan determines how you are going to produce your product or deliver services to the public. It includes elements such as purchasing, delivery, inventory, equipment, personnel, operation flow, and building design.

3. *Financial plan:*   This plan considers how you are going to operate your business, reduce losses, and make a profit.

4. *Record-keeping plan:*   This plan establishes procedures that account for all stages of business activities for the purposes of payroll, benefits, taxes, insurance, and compliance with laws and regulations.

Incorporating waste reduction into your business plans may be the most useful form of waste reduction because of the ability to address waste generation before waste-producing materials, processes, and procedures can enter and become established in the business. Moreover, curbing waste production through business planning typically requires little money or time to implement.

These planning considerations are especially important for new businesses. By considering waste reduction early on, a company can avoid the losses that have been experienced by other businesses.

7.2 Waste Reduction and Your Business Plan

Here are some methods for promoting waste reduction in the various parts of a business plan. This chapter does not discuss all such options, but should provide enough guidance to initiate a thorough examination. The Business Plan Evaluation Checklist (Fig. 7.1 in Sec. 7.3) and the list of resources in Sec. 7.4 provide additional information and advice.

7.2.1 Waste reduction in the market plan

Successful business managers know what goods and services their customers want. They also recognize who else in the community is supplying the same goods or services, i.e., their competitors. Business managers will often use this knowledge to develop marketing strategies for their company. In a competitive economic climate, consideration of waste reduction in your marketing plan may help you maintain a competitive edge over other businesses providing similar services.

Throughout the nation there is widespread support for prevention of environmental pollution, particularly at the local level. A company actively promoting waste reduction has a good chance of receiving strong support from local communities and the press. The money saved and made from a waste reduction program can be used elsewhere in the company, which might give a business that extra competitive edge in the marketplace.

Many companies are located in states or localities where they are eligible for so-called Pollution Prevention Awards. Typically these awards are given annually to businesses actively involved in preventing pollution. A company winning this award generally receives free publicity (in a newspaper press release and television news features) and is given a plaque to display. This type of free publicity should not be overlooked in a marketing plan.

7.2.2 Waste reduction in the operation plan

Policies for the operation and design of a business in addition to procedures for purchasing, inventory, delivery, and personnel can ultimately provide the focal point for waste reduction in your shop. Provided below are techniques for incorporating waste reduction into a business operation plan.

Process operations.   The industrial processes in your business are typically the major source of waste production and thus provide enormous opportunities for reducing waste. Maintenance work, tank cleaning, degreasing, painting, printing, and other processes can be designed to produce the least possible amount of waste.

When making decisions on the type of processes to use in a business and in maintaining existing equipment a business manager or waste reduction audit team should consider source reduction, material recovery, equipment design, waste segregation, maintenance programs, and employee education.

Source reduction.   Design or redesign the process to control waste production at the source or to recover any wasted raw materials. For example, dry-cleaning processes can be designed to trap solvent emissions and return the solvent to the cleaning process.

Material recovery.   Modify process equipment and operations to enhance recovery or recycling of process materials.

Equipment design.   Redesign equipment or production process to reduce waste.

Waste segregation.   Segregate waste to enhance recycling (both on site and off site) and waste exchange. One company's waste may help fuel the operation of another business.

Maintenance programs.   Initiate a preventive maintenance program that ensures efficient operation of equipment and handling of raw materials, eliminates leaks and spills, and monitors for additional waste reduction opportunities.

Employee education.   Emphasize employee development and training as a vehicle for promoting waste reduction and maintaining efficiency.

Purchasing procedures.   Changes in purchasing procedures can control potential losses before they enter the shop. When reviewing company purchasing procedures, consider prepurchase review, inventory control, equipment evaluation, centralized purchasing, material safety data sheets (MSDSs), property survey, facility design, and expansion plans.

Prepurchase review.   Evaluate materials prior to purchase to ensure that a potential supplier is using nontoxic (or least toxic) materials that are the least costly to handle. State waste reduction programs can provide assistance in evaluating material purchases.

Inventory control.   Control chemical inventories to prevent possible spills and overpurchasing (thus preventing the resulting costs for disposal of extra materials). Buy only what you need.

Equipment evaluation.   Purchase only equipment that is predesigned to reduce waste and comply with applicable environmental and occupational safety and health standards. State waste reduction programs, equipment suppliers, and distributors may be able to provide some of this information.

Centralized purchasing.   Purchase materials through a central person or department in your business to eliminate all unnecessary purchases and ensure that all waste reduction purchase policies are followed.

MSDSs.   Make certain that distributors supply MSDSs when you purchase hazardous materials. MSDSs are chemical information sheets that can help business managers determine the nature of their hazardous materials and potential wastes. Moreover, these sheets will help owners comply with state and federal worker and community right-to-know laws.

Property survey.   When buying real estate, require the seller to pay for an independent survey documenting that the property is free of any hazardous materials contamination and/or abandoned wastes (such as asbestos or PCBs). Owning or leasing the property will potentially make you responsible (liable) for any cleanup costs.

Facility design.   Evaluate potential building purchases to determine whether or not the building design is amenable to waste reduction. For example, does the building design enable shop equipment to be laid out in an efficient manner?

Expansion plans.   Always consider waste reduction when making plans for business expansion.

Receiving procedures. How often have you ordered supplies and had to return them because they were the wrong product? How often do you have to clean up spills resulting from improper shipment and packaging?

Once you've made a purchase, there are several simple methods for eliminating waste generation in the receiving component of your operation plan. You should designate a receiving area, train receiving employees, use quality suppliers, review purchase agreements carefully, and document your agreements with suppliers.

Designate a receiving area.   Your operations plan should designate a specific area for receiving all materials that enter your company. This will allow you to control materials as they enter. A receiving area can also be designed to prevent and control wastes from spills, leaks, and so forth.

Train employees.   Employees should be trained to handle shipments properly to prevent property losses, injuries, and costly waste disposal. They should also be knowledgeable about emergency procedures in the event of a spill or injury.

Use quality suppliers.   Don't buy from just anybody. While the lowest bid is important, the quality and reliability of the supplier is equally crucial. It may be worth paying a little more for a supplier's products if the supplier delivers those goods intact and according to specifications. Check the track record of the supplier to determine if its performance is of the highest quality and integrity. Request references from the supplier.

Review purchase agreements.   Purchase order agreements should include terms and conditions for receiving material orders. These conditions could include provisions allowing you to inspect order materials prior to acceptance. If containers are leaking when your order arrives, for example, the terms could state that the supplier or shipper bears the responsibility for cleanup or whatever out-of-pocket expenses you must pay to control the incident and protect your employees.

Document agreements.   All terms should be documented to ensure that all conditions are followed according to specifications. Suppliers are in business to provide materials to you. They want your business and should be willing to accept shipments on your terms. If a supplier is not willing to do this, find another supplier or change products if you can.

Delivery procedures.   The same methods for controlling waste during receiving can be applied to the delivery aspect of your business plan. When making deliveries, special delivery arrangements might also include designated receiving areas, trained material handlers, delivery agreements, and documentation.

Designated receiving areas.   Have customers designate a single area for receiving shipments, with measures to reduce accidents and spills.

Trained material handlers.   Request that customers use only trained workers for hazardous materials shipments at the receiving dock. This policy will protect both companies from future liabilities. It will also protect the health and safety of employees and the surrounding public.

Delivery agreements. Require that the customer inspect the delivery and document that the materials have been received in an acceptable condition. This will help reduce any potential liabilities once the goods have been received by the customer.

Documentation. Get a copy, in writing, of all delivery agreements to ensure that proper procedures are followed.

Inventory procedures. An inventory policy is an important component of a business plan. How much extra stock should be kept on hand? This question and others are difficult to answer, particularly for businesses operating in rural areas, away from major transportation routes. Businesses need enough inventory to allow smooth operations until the next order is received. However, too much inventory may result in the production of waste and in increased financial hardship on a business (by tying up spare cash). Waste reduction should be a prime consideration when developing the inventory component of a business plan. When developing or reviewing such plans, consider issues such as materials inventory, material shelf life, emergency controls, container size, storage areas, and procedural changes.

Materials inventory. Materials inventory presents special concerns for any business. The more inventory of reactive, toxic, or ignitable materials you have, the greater the chances for exposure to workers and safety problems such as spills and fires. The downtime and other problems associated with these losses will potentally outweigh the costs of waiting for a resupply shipment.

Material shelf life. What is the shelf life of the material that you intend to purchase? Leftover inventory with a limited shelf life will have to be disposed of or sent back to the supplier at company expense. Moreover, simply leaving outdated stock in the inventory may overcrowd inventory space, increasing the potential for fire, worker exposures, and regulatory violations.

Emergency controls. Decisions concerning the amount of hazardous materials inventory to be kept in a shop should be preceded by an evaluation of emergency control systems. Can fire suppression systems handle large inventories? Are storage areas designed to handle spills and other mishaps associated with large chemical inventories? Failure to consider emergency controls may result in increased waste generation and financial and human losses.

Container size. Packaging may have a direct bearing on the amount of waste a company can reduce. Chemical inventories stored in large containers (55-gallon drums) may result in increased waste generation when compared to the same chemical stored in smaller containers. For example, chemicals stored in large containers typically have to be transferred to smaller ones before use, which can result in spills, material evaporation, and batch contamination if the material is returned to the same drum. Purchasing materials in appropriate-sized containers can save money and promote more efficient use and handling of chemical materials.

Storage areas.   Always consider the area where materials are stored in a company. Is the area designed to control spills? Are incompatible chemicals stored next to each other? (Reactive chemicals should not be stored next to each other.) Does the storage area take into account potential damage from earthquakes?

Procedural changes.   Company owners may leap at the chance to buy a huge inventory of materials offered at a discount. It makes good financial sense to buy the most product for the lowest rate, right? But think a moment—is the product really needed? Future process changes may not require as much or any of that material. When that happens, a company has to bear the cost for disposing of the material or finding another company to use it, not to mention the original purchase outlay.

Personnel policy.   The success of a waste reduction program depends, in part, on company employees. Workers must be trained to handle raw materials properly and carry out the company waste reduction program. Personnel policies should include ample opportunity for staff development and training in this regard. With adequate training and education employees will handle materials properly and reduce occupational exposures, while at the same time supporting company waste reduction efforts. Moreover, workers are often in the best position to recommend process changes that may enhance waste reduction in the shop, because they are on the front line of the business operations. Company managers should encourage workers to make suggestions on how to reduce waste, thereby improving overall business efficiency.

7.2.3   Waste reduction in the financial plan

Creating a sound financial plan is an important element in any successful business venture. Maintaining a profit can be especially difficult for a small-business owner. Even seemingly small losses will add up year after year and can determine whether a business prospers or fails. Therefore, it is essential to include waste generation and reduction as a cost of doing business. Profit and loss statements will be inaccurate if they do not include waste management expenses or any waste reduction savings.

This type of planning requires a business to maintain accurate figures on the types and quantities of industrial wastes generated, along with their associated management costs. These costs include waste disposal, transportation, management overhead, regulatory compliance, insurance, and so forth. What may seem like a cost-efficient waste management approach may not be such when it is considered within the context of a company's overall financial picture.

Thus, waste management cost information will help a business develop more accurate profit and loss statements and create a more realistic financial plan. In addition, this information is crucial when evaluating the cost-effectiveness of any planned or existing waste reduction measures.

7.2.4 Waste reduction in the record-keeping plan

Keeping accurate records of business operations is not an easy chore, particularly for a small business, which may not have the resources to handle large amounts of paperwork. However, sound recordkeeping is important in maintaining a business plan and thus is an important step in setting up a waste reduction program.

Keeping accurate records on the types and quantities of waste produced is important for determining how they can be reduced and by how much. With this information you can calculate material losses in process from spills, system leaks, and other poor handling. Moreover, these records will allow you to evaluate the economic and technical effectiveness of the various elements in a waste reduction program. In addition, attention to paperwork and records will help document waste management program activities and thus help companies with complex federal, state, and local regulations.

As mentioned previously, maintaining MSDSs is necessary for compliance with worker and community right-to-know laws. Reviewing these data sheets will also help to identify potential hazardous wastestreams. This can be a low-cost method for initially identifying wastes that can be reduced or eliminated. Product manufacturers and distributors are required to provide companies with MDSDs, which in turn need to be available to employees.

7.3 Evaluating a Business Plan

Figure 7.1, the Business Plan Evaluation Checklist, is designed to help interested readers evaluate a business plan with a critical eye toward waste reduction. The checklist can also be used as a guide for developing companies that want to integrate waste reduction into their business plans and procedures.

Figure 7.1 is divided into four parts, which correspond to the major categories of a business plan previously covered in this chapter. A series of questions is provided with possible *Yes, No,* or *Not sure* responses. A *Yes* response means that business plans are promoting or could promote waste reduction. A *No* response means that business plans are not encouraging waste reduction. A *Not sure* response means that the business plan (related to the topic in question) needs more evaluation to determine if it encourages waste reduction. The checklist should be saved upon completion and used periodically to update a business plan and to document areas where a business plan needs to be changed to promote waste reduction.

7.4   Resources for Further Assistance

For more information about how to develop a business plan and how to make sure it includes wastes reduction, refer to the following resources:

- Environmental Protection Agency, "Waste Minimization—Environmental Quality with Economic Benefits," Office of Solid Waste and Emergency Response, Washington, D.C., 1987.
- Center for Hazardous Materials Research, "Hazardous Waste Minimization Manual for Small Quantity Generators," University of Pittsburgh Applied Research Center, Pittsburgh, Pa., 1987.
- Monica E. Campbell and William M. Glenn, "Profit from Pollution Prevention," Pollution Probe Foundation, Toronto, Ontario, Canada, 1982.
- The National Roundtable for State Waste Reduction Programs, c/o The Department of Natural Resources Pollution Prevention Pays Program, P.O. Box 27687, Raleigh, N.C. 27611-7687.
- Any state small business development center.
- Alaska Health Project Waste Reduction Assistance Program (WRAP), 431 West 7th Avenue, Suite 101, Anchorage, Alaska 99501, (907) 276-2864.

This chapter was adapted from David Wigglesworth's "Profiting from Waste Reduction in Your Small Business," Alaska Health Project, Anchorage, Alaska, 1988, pages 9–18 (Produced with support from the Charles Stewart Mott Foundation).

Figure 7.1 Business plan evaluation checklist.

Company: _____

Date completed: _____

Person completing form: _____

INSTRUCTIONS:
- *Check* the appropriate answer for each question.
- *Yes* means that the company business plans are promoting waste reduction.
- *No* means that the company business plans are not promoting waste reduction.
- *Not Sure* means that the company needs to further evaluate the plan in that area.

	Yes	No	Not Sure
A. MARKET PLAN EVALUATION			
1. Do you and your workers recognize the importance of proper materials management and waste reduction?	[□]	[□]	[□]
2. Do your marketing strategies incorporate the positive public image related to waste reduction?	[□]	[□]	[□]
3. Do you publicize your company's efforts to reduce waste?	[□]	[□]	[□]
B. OPERATIONS PLAN EVALUATION			
1. Are workers and management developing a plan to promote waste reduction in your company?	[□]	[□]	[□]
2. Are business procedures designed to promote source reduction?	[□]	[□]	[□]
3. Is the company recycling every waste that it can?	[□]	[□]	[□]
4. Does the company know the quantity of waste (liquid, solid, gaseous) produced by each process?	[□]	[□]	[□]
5. Is the company shop kept clean and orderly to reduce the chance of spills and increase efficiency?	[□]	[□]	[□]
6. Does the company have a recycling program for computer, ledger, and mixed paper?	[□]	[□]	[□]
7. Has the company determined that air emission waste produced in the plant can or cannot be reclaimed?	[□]	[□]	[□]
8. Are process wastes segregated to enhance recovery of raw materials?	[□]	[□]	[□]
9. Do workers know which process reduction produces wastes?	[□]	[□]	[□]
10. Does the company operations plan include periodic waste reduction audits?	[□]	[□]	[□]
11. Does your company maintain MSDSs to evaluate raw materials prior to purchase to ensure you are using the least toxic material possible?	[□]	[□]	[□]

(Continued)

	Yes	No	Not Sure
12. Is inventory stock limited to prevent possible spills and to avoid overpurchasing and other waste?	[□]	[□]	[□]
13. Does the company request prepurchase information on the waste generating potential for new equipment?	[□]	[□]	[□]
14. Do purchase agreements include provisions for inspecting shipments of raw materials prior to acceptance to ensure that they are not leaking and creating a waste?	[□]	[□]	[□]
15. Does the company attempt to exchange wastes that cannot be reduced at the source?	[□]	[□]	[□]
16. Can fire suppression systems handle a major emergency involving the hazardous material used and waste produced?	[□]	[□]	[□]
17. Are storage areas designed to minimize earthquake damage, control spills, and other mishaps?	[□]	[□]	[□]
18. Are all workers trained about what to do in the event of a hazardous materials incident?	[□]	[□]	[□]
19. Does company policy promote employee training and development in the area of waste reduction?	[□]	[□]	[□]

C. FINANCIAL PLAN EVALUATION

	Yes	No	Not Sure
1. Are waste generation and disposal costs included in profit and loss statements?	[□]	[□]	[□]
2. Have waste generation costs been determined for each process in the business?	[□]	[□]	[□]
3. If yes, do you "charge" the costs directly to the process?	[□]	[□]	[□]

D. RECORD-KEEPING PLAN EVALUATION

	Yes	No	Not Sure
1. Are records maintained on the amount of raw material used per process to monitor process efficiency?	[□]	[□]	[□]
2. Are logs maintained on the types and the quantities of waste produced so that specific wastes can be targeted for waste reduction?	[□]	[□]	[□]
3. Are MSDSs or equivalent information used to help identify possible hazardous waste-stream?	[□]	[□]	[□]
4. Does the company have written plans and procedures to document plant operation procedures and waste reduction policies?	[□]	[□]	[□]

Incorporating Waste Minimization into Research and Process Development Activities

Steven C. Rice, P.E.

BASF Corporation
Parsippany, New Jersey

Waste minimization is becoming an increasingly important aspect of virtually all industrial organizations, large and small. Real incentives for reductions are provided by disposal cost increases on the order of 25 to 50 percent per year and by potential long-term waste generator liabilities associated with waste disposal site cleanups. When combined with regulatory pressures, shrinking disposal options, and minimization certifications, there are significant driving forces to reduce the volume and hazardous nature of our wastes.

This chapter presents practical experiences with waste minimization for research and process development activities and describes the difficulties that appear to be rather common among those organizations which have attempted to reduce their wastes. The chapter discusses the unique characteristics of experimental units, waste reduction opportunities and experiences, the tracking and reporting of such efforts, and the future challenge. The information presented reflects a compilation of personal experience, as well as the experience of my counterparts in other organizations, based on our mutual contacts, activities, and discussions. Thus, individual details should not be considered to be automatically applicable to all companies or organizations, as each organization will have its own situations for which success or failure may be realized.

8.1 Unique Characteristics of Experimental Units

Recent conference presentations, journal articles, and an Office of Technology Assessment report[1] have provided a significant amount of information on the economics, technologies, and operational changes for waste reduction. Unfortunately, only a small amount of this information has applicability to the research and process development environment[2,3,4] because of the unique characteristics of experimental units. The three unique characteristics of experimental units—diversity, variability, and originality—all suggest that the waste reduction opportunities for experimental activities may be quite different from reduction opportunities that are applicable to manufacturing operations.

8.1.1 Diversity

Process development groups are always working on a wide range of projects in diverse interest areas. Approaches suitable in one area with one set of constraints may be quite unsuitable for another area.

8.1.2 Variability

The nature of research work requires that ever-changing techniques and activities be attempted, which is very unlike operating plants. A problem once resolved for one experiment may no longer be resolved a few days or weeks later, since the work, the kinds of materials, and the amounts of materials may have changed. New circumstances are constantly being presented as new work is initiated. Therefore, waste reduction for research and process development activity is much more than a "one-shot deal" to implement. It becomes more of an effort to create a mental attitude within each research scientist or process engineer to think of waste reduction continually in all phases of his or her work.

8.1.3 Originality

One of the many purposes of process development is to do new work, often in new areas, so a researcher can rarely go to published literature to find out how someone else resolved a particular environmental issue. Also, a project's environmental impact and waste generation characteristics may not always be clear at the beginning of the effort.

8.2 Waste Reduction Opportunities and Experiences

Waste reduction opportunities are as varied as the nature of the research work. Most can be categorized into one of the following approaches:

- Provide research employees with information
- Purchase smaller-sized units of stock chemicals
- Restock unopened materials
- Create an internal material exchange
- Allocate disposal costs
- Treat at the source
- Conduct environmental reviews

The advantages and successes of these various approaches will depend on the specific organization and site. Each approach is discussed below.

8.2.1 Provide employees with information

Because of the diverse, semiautonomous climate in which most research is conducted, development of an organizational mind-set becomes one of the first and most important waste reduction opportunities to pursue. Without such a mind-set, further efforts may have little or no effect if there are not constant reminders and reviews. Through a series of internal seminars and presentations, possibly coupled with letters and brochures, employees can be provided with detailed information on why reduction is important, the cost-saving potential, and the basic rationale for such a program.

This information should review the methods discussed below, yet also encourage attempts at innovative approaches, ideas, and solutions. In order to garner the maximum amount of interest and support for such an effort, it is important to emphasize "what's in it" for the individual. Approaches which have met with success are ones that suggest (1) that waste reduction means reduced operating costs, thus freeing money from the budget for other work in the individual's area and (2) that new or modified processes for the competitive advantage of the organization can be developed. This second aspect will be discussed in more detail later in the chapter.

Vital to the ultimate success of this opportunity is support of executive management for the program. Without such support and commitment, any efforts are likely to yield only marginal results.

8.2.2 Purchase smaller-sized units of stock chemicals

The stocking of smaller-sized units of stock chemicals is more of a materials handling consideration than a technical consideration. This approach involves making available the smallest-sized units of the chemical necessary for an experiment. Instead of stocking only gallon-sized containers of a material, consider stocking additional quantities of quart- or pint-sized containers.

While the initial cost for the chemical purchase may be higher, in many cases disposal costs for unwanted remainders often exceed the purchase cost. Thus, the increased purchase cost will be more than offset by the reduced waste cost. Of course, in certain instances larger unit sizes are truly needed and justifiable, and these can be handled on a case-by-case basis, rather than through the stockroom.

8.2.3 Restock unopened materials

A second method for keeping nonwaste from becoming waste is to restock unopened stock chemicals. Research scientists and engineers who possess unopened stock chemicals should be encouraged to return these materials to the stock room instead of discarding them. In this way, others may be able to use the materials which might otherwise become wastes. Each container may have to be dated to determine if its useful shelf life has been exceeded.

The restocking of opened stock chemicals is also an alternative. However, in research work this is seldom practical due to the high-quality material needed and by each researcher's uncertainty about the quality of the material remaining in an opened container.

8.2.4 Create an internal materials exchange

Perhaps the largest opportunity for keeping nonwaste from becoming waste exists in the creation of an internal materials exchange or "classified ads" system to keep surplus nonstock or open-stock materials from being discarded unnecessarily. This can be done in a variety of ways; for example, an organization might develop a multi-accessible computer network or organize a manual system utilizing a trained contact person.

Regardless of the type of information system used, it is important to refer to the materials by their chemical abstract number due to the various synonyms used for the same material. Also, each system should have "available" as well as "wanted" listings which contain information such as the person owning or wanting the material, the location, the amount and grade of material, the size of the container, and the date when it is desired or will be available. It would be important to purge the entry information periodically.

8.2.5 Allocate disposal costs

Allocation of disposal costs is perhaps the most effective tool that can be used to heighten individual awareness of waste reduction. Often the researcher and the organization is not aware of the costs of disposing the wastes from a subject research. One very effective way to encourage such consideration is the allocation of waste disposal and handling costs to a separate budget line item, internally billed directly to the account of the generating project or organizational unit. When such costs show up as an accountable expenditure which the generator must budget and track throughout the year, additional attention is directed to them. In some situations a revised accounting system may be required for such separate billing.

Essential to the viability of this opportunity is the accurate identification of the generating lab or project on each container's waste identification form. The disposal and handling cost of each container of waste then can be back-charged to the generator as part of his or her total project budget. Experience has shown that in a research environment frequently the individual cost of each container cannot be determined. In those cases a prorated share of the disposal costs for similar materials can be distributed among the generators.

8.2.6 Treat at the source

Treatment at the source has been emphasized in most of the published information applicable to research and development (R&D) activities.[5,6] While there are some advantages to this approach, there is a significant caution which should be given careful consideration before proceeding: there may be regulatory implications with "treating" hazardous wastes without an appropriate permit. Authorities such as the U.S. Environmental Protection Agency (EPA), the applicable state agency, and possibly an environmental specialist should be consulted prior to pursuing this activity.

8.2.7 Conduct environmental reviews

An environmental review program for process designs can provide the opportunity for necessary waste reduction issues to be considered during the project or unit design phase, prior to construction. There are several ways to address this opportunity, the specifics of which need to be formulated to meet the needs and objectives of the individual organization. One technique incorporates the project review within the regular planning and budget process. Figure 8.1 shows how a review system, which includes safety, environmental, and industrial hygiene aspects, can be structured. A copy of the environmental portion of a review form is presented in Fig. 8.2.

Note that there are two items specifically relating to hazardous waste generation and minimization. If nothing else, the opportunity to respond to the requested information may get the process designer to think about the process in a slightly different light. Responses such as "according to company procedures" or "as usual" usually indicate that the respondent doesn't know. Further follow-up by the site environmental contact may be necessary.

A key advantage to a waste minimization review at this stage of the research work is that such consideration can be addressed at a time when cost-effective solutions can be incorporated easily. For example, in one organization a unit was proposed for the separation of rubber polymer from a solvent base. The original design had a simple stripper which vaporized and vented much of the solvent, with the remaining wet residual polymer to be handled as a hazardous waste liquid. After considering waste disposal, designers were able to make minor modifications that upgraded the wastestream so that virtually 100 percent of the polymer could be recovered for reuse and almost all of the solvent could be recycled back into the process. Not only did this reduce hazardous waste generation, but those design improvements had a payback time of only about two months due to reduced waste disposal and feed solvent costs.

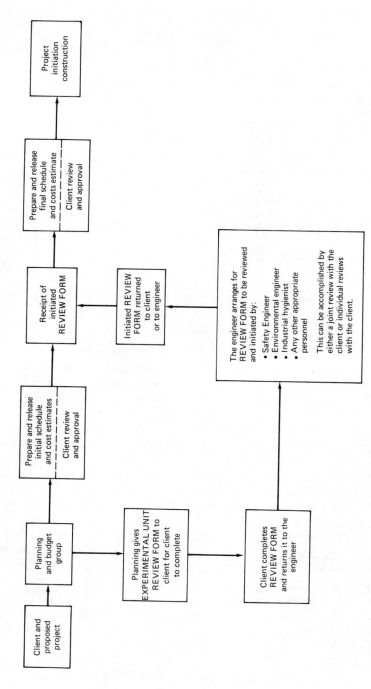

Figure 8.1 Flow plan for experimental unit review form.

F. Environmental concerns

Will there be any potential for	Yes/No	Material(s)	Maximum emission rate per hour	per day
• Air pollutant emissions?	_____	_____	_____ (lb)	_____ (lb)
(includes use of vent lines)		_____	_____	_____
• Wastewater discharge?	_____	_____	_____ (gal)	_____ (gal)
(includes overflow or release to sewer)		_____	_____	_____
• Ground/soil contamination?	_____	_____	_____	_____
(includes spill during operation or material handling)		_____	_____	_____

Are there any emission control devices anticipated in the unit's design?

☐ exhaust filters   ☐ cyclones   ☐ scrubbers   ☐ liquid/vapor separators   ☐ other emission control devices?

If so, describe:_____

Describe how wastes (unused feeds, samples, and products) will be identified and handled.
_____

Describe how the volume or hazardous nature of the wastes can be reduced.
_____

Figure 8.2   Environmental section of experimental review form.

8.3   Tracking and Reporting of Waste Reduction Efforts

A key element to any serious waste reduction effort is the tracking of how well a program is doing. This is usually done by comparing waste generation quantities and costs from one time period to another, with figures often normalized to a common factor such as throughput, production, or sales. Reports are developed for internal evaluation and submitted to regulatory authorities pursuant to applicable regulations.

The tracking and reporting of waste reduction efforts by research and process development poses unique challenges due to the inherent variability of these types of activities. First, project initiation or expiration impacts the types and quantities of waste generated each year. For example, the initiation of a single large unit can overshadow current waste generation quantities and wipe out true reduction successes. Conversely, an expiration of one or two key projects can inaccurately show up as a significant waste reduction success.

Second, laboratory cleanouts can also distort an individual year's generation, especially in a smaller organization or if cleanouts are not conducted frequently. It is not uncommon to see large increases in annual generation due solely to a sitewide cleanout.

These challenges can be addressed through the selection of a proper measurement quantity and the use of a suitable normalization factor.

Proper quantification of the waste is a vital precursor to accurate tracking and reporting. Usually the manifested shipping weight is used as the basis for the amount of waste generated. However, unlike in a manufacturing plant, wastes from research or development activities are often accumulated in a "lab pack," which is a drum or box of several individual but compatible wastes in containers (see Fig. 8.3). In such a case the glass or metal containers, cushioning material, and metal drum can make up the majority of the total weight. Thus, the actual amount of waste material is difficult if not impossible to determine.

Converting to a mixture of weights and volumes to report the various wastestreams can alleviate this problem somewhat, particularly if the much heavier, drummed liquid wastes are reported separately from lab-pack wastes. The trade-off is that the routine administrative tracking and reporting then becomes more cumbersome as (1) most wastes may then require *two* reporting parameters and (2) many lab-pack liquids may need to be manifested, shipped, and reported under the Department of Transportation (DOT) "solids" designation. Alternately, the use of a DOT-approved "80-lb" cardboard box (usually used for incinerable materials) can contain a much higher weight percent-

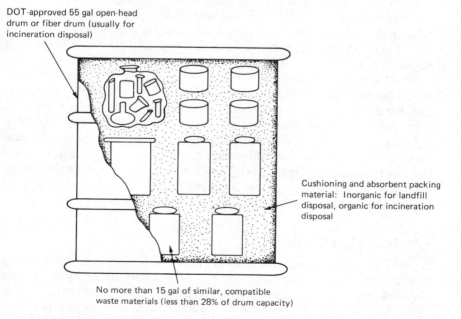

DOT-approved 55 gal open-head drum or fiber drum (usually for incineration disposal)

Cushioning and absorbent packing material: Inorganic for landfill disposal, organic for incineration disposal

No more than 15 gal of similar, compatible waste materials (less than 28% of drum capacity)

Note: DOT-approved boxes (for incineration) can contain a much higher proportion of waste volume to container capacity

Figure 8.3 Diagram of a "lab pack."

age of waste, minimizing the inaccuracy of a weight-based measurement approach. It is clear that each organization should investigate the associated advantages and disadvantages in light of its own particular wastestreams and set of circumstances.

The use of a normalization factor in reporting waste generation and reduction is instrumental in generating useful data for internal or regulatory agency study. The state of New Jersey recognized this several years ago and incorporated the concept into its annual reporting requirements for hazardous waste generators. This raised the question of what would be an appropriate factor for the R&D industry? Factors such as production and sales were clearly inappropriate for the subject activities.

Several factors such as research dollars, number of research units, and number of researchers, among others, are somewhat indicative of research activity and should have at least a casual relationship to waste generation. While research dollars may be the most accurate normalizing factor, often it is proprietary information and not available for release. In some cases a factor such as an expenditure "index" may be appropriate and can provide confidentiality as long as the index basis is secure.

In general, number of units is not a useful or desirable factor. Units can be operating or idle at almost any given time or for a duration. Also, not all units are the same, for example, a wet chemistry unit would not be comparable in waste generation to a developmental pilot plant.

Thus, the total number of on-site people appears to be a suitable, albeit not perfect normalizing factor. For intensive research sites, this number should be close to the total research population and somewhat indicative of site activity and waste generation. For less intensive office-research sites, this number would be less sensitive to annual swings and variations in individual research programs. Keep in mind that as the resultant waste per person will tend to be much lower at this second type of site, there is an inherent difficulty in comparing waste generation rates between the two types of sites.

8.4   The Future Challenge

A few short years ago the first industrial waste reduction efforts were directed toward alternate disposal strategies, usually away from land disposal and final treatment or incineration. Lately, the thrusts have been directed toward changing in-plant procedures, equipment, or feedstocks. The future direction for industry may be to go further upstream, that is, to address waste reduction before a plant or process is designed and constructed.

Thus, as an integral part of the total waste reduction effort, industry faces a rather formidable future challenge: to develop new or modified processes which incorporate waste reduction into the processes' inherent designs. This approach probably has the potential to provide the most significant, long-term reduction in industrial waste generation by ensuring that the amount or hazardous nature of any waste is reduced to the minimum level possible, perhaps even to elimination. Through new or modified designs, the concept of how to reduce wastes can be included within the chemistry and engineering of the process. The review technique discussed earlier can be an instrumental tool in addressing this challenge.

An effort such as this has advantages which may be more clearly understood by others less familiar with hazardous waste issues if put into terms that are more familiar to them. First, such an effort has the potential to result in increased economic advantage for the company or organization. Waste reduction of 10 percent, when put on a large-scale commercial basis, yields substantial savings on waste disposal costs. When that 10 percent is compounded by significant annual disposal cost increases for research-type wastes, the potential rate of return on investment becomes even greater.

Additionally, these reductions can change what might otherwise be a noncompetitive product improvement into a highly competitive one. A process unit with a marginal (or negative) financial analysis can show substantial profit when the future waste disposal cost savings are included in the evaluation. This aspect is especially important in mature industries, where significant industrywide process improvements may not otherwise be available.

Recognition for both personnel and the organization as a whole can come from a waste minimization effort. Individuals can be recognized for contributing to the success of the organization. Patents, awards, and promotions are but a few of the possible rewards. For the organization, such an effort can make the organization's work valuable to the commercial sectors it supports, which in turn can bring more (or better-supported) work into the organization. Of course, there are also the public affairs aspects, which can emphasize waste minimization efforts to promote the favorable public image every organization strives to achieve.

8.5 Summary

The variety in experimental units requires unique approaches to waste minimization. Approaches common to manufacturing operations are seldom applicable or effective for research and process development activities. Nevertheless, there are several waste reduction opportunities available, many of which are relatively simple and inexpensive to initiate. An environmental review during a project's design phase can provide a forum to discuss waste reduction.

Waste reduction tracking and reporting are difficult, due to the nature of R&D activities and the types of wastes they generate. Accurate, meaningful comparisons between different organizations or across time frames require more than a cursory investigation. The future challenge is to develop new or modified processes which incorporate waste reduction within the processes' inherent designs. Besides long-term waste reduction, increased organizational competitiveness and personal recognition are but two of the rewards which can result from such an effort.

8.6   REFERENCES

1. Congress of the United States, Office of Technology Assessment, "Serious Reduction of Hazardous Waste: For Pollution Prevention and Industrial Efficiency," OTA-ITE-317, U.S. Government Printing Office, September 1986.
2. S. C. Rice, "Environmental Review Strategy for R&D Activities," *Environmental Progress*, February 1988, pp. 46–51.
3. E. A. Holland, "The R&D Sector: Optimizing Waste Minimization Practices," NJDEP/USEPA Hazardous Waste Reduction Audit Workshop, November 17, 1987.
4. W. C. Hollinsed and S. T. Ketchen, "Waste Reduction Through Minimization of Reagent Usage," *Hazardous Waste and Hazardous Materials*, November 1987, pp. 357–361.
5. American Chemical Society, "Less is Better—Laboratory Chemical Management for Waste Reduction," 1985.
6. R. A. Field, "Management Strategies and Technologies for the Minimization of Chemical Wastes from Laboratories," Duke University Medical Center Division of Environmental Safety/North Carolina Pollution Prevention Pays Program, September 1986.

Waste Minimization for Small Quantity Generators

Scott M. Raymond

President, Raymond Environmental Compliance Services
Pittsburgh, Pennsylvania

9.1 How Can Small Quantity Generators Apply Waste Minimization?

An important part of any hazardous waste management program is waste minimization. Companies that generate small quantities of hazardous waste (called *small quantity generators* or SQGs) can reduce or eliminate many regulatory requirements, waste-handling expenses, and other liabilities by implementing simple, low-cost waste reduction alternatives. Numerous case studies have shown that reducing the amount of waste generated results in simultaneous economic and environmental benefits, regardless of a company's size.

Waste minimization is good business. Recycling, material substitution, good housekeeping, and replacement of outdated equipment are all sound business practices that also reduce quantities of hazardous wastes generated. Waste minimization activities can help lower operating costs, save on raw material expenses, and improve employee safety performance.

SQGs can reduce from 15 to 100 percent of the hazardous wastes they must manage by implementing various waste minimization techniques. It is recommended that SQGs begin with simple, low-cost approaches, such as good housekeeping and improved material management, and then progress to more difficult and expensive methods, such as installing recycling systems or equipment modification. Chapter 3 provides some details on waste minimization techniques and technologies.

Start by identifying where you need waste minimization. Develop an inventory of the hazardous wastes your company generates by conducting a waste minimization assessment for your operation. Specific procedures for a waste minimization assessment should be tailored to your business. However, all waste minimization assessments should have the following steps:

- Review of background and files
- Visual inspection of all operations
- Listing of waste minimization opportunities

See Chapter 5 for a discussion on the procedures for conducting a waste minimization assessment.

The following sections provide checklists of waste minimization options that can be implemented in specific industrial situations. These can be used when developing a list of waste minimization opportunities for your company. Twelve industrial categories of SQGs are addressed. Waste minimization alternatives for your industry, as well as those for related industries, should be considered when developing a list of options for your company.

9.2   Construction

Construction is recognized as a major generator of hazardous wastes among SQGs. The industry group includes a large number of professions and operations, which produce numerous kinds of wastes in variable volumes. The most common hazardous waste types generated by the construction industry are listed below:

- Spent solvents
- Strong acids or alkalies
- Paint wastes with heavy metals
- Ignitable wastes
- Vehicle maintenance wastes

Wastes of different types should be segregated. Separating hazardous wastes from nonhazardous wastes can decrease volumes that must be specially handled. Segregation also keeps wastes from becoming contaminated so they can be recycled easily. However, some or all of the wastes within a single group may be combined before treating or disposing, as long as the wastes are compatible.

Table 9.1 provides a checklist of waste minimization alternatives that have been successfully used in the construction industry. Additional waste minimization options may be found in other sections of this chapter which address industry groups that may be related to the construction industry.

9.3   Dry Cleaning/Laundries

Dry cleaners that use hazardous solvents, as well as most industrial laundries, generate some hazardous wastes. Examples of hazardous wastes generated by dry cleaning operations and laundries are listed below:

- Still residues from solvent distillation units
- Spent filter cartridges
- Cooked powder residues
- Oil and grease sludge contaminated with heavy metals

Most waste minimization practices for dry cleaners and laundries are centered around good operating practices, housekeeping, and recycling. Use of procedural measures that extend the life of solvents, loss-prevention practices, and wastestream segregation are all waste reduction options that can be applied. Table 9.2 provides a checklist of alternatives that can be used by the dry cleaning and laundry industries.

TABLE 9.1   Checklist of Waste Minimization Alternatives for the Construction Industry

Waste category	Alternatives
Spent solvents and cleaners	Avoid cross-contamination of solvent. Avoid water contamination of solvent. Remove sludge continuously. Keep solvent containers covered to reduce evaporation. Monitor solvent composition. Consolidate cold cleaning operations. Use mechanical blasting for paint stripping. Use nonhazardous or nonchlorinated solvents, where possible. Use a solvent recycling service. Consider using on-site solvent distillation units.
Acid/alkaline cleaners	Improve operating practices. Avoid spillage and dripping of acid/alkaline cleaning solutions. Segregate acid and alkaline wastes. Substitute nonhazardous or less hazardous materials. Remove sludge from acid or alkaline baths frequently. Use mechanical cleaning methods, where feasible. Avoid cross-contamination. Reuse by filtering and rejuvenating.
Paint wastes	Reduce use of solvent and metals-based paints, where possible. Use water-based coatings. Use high-solids coatings, such as powder coatings. Implement better operating practices for painting. Segregate all wastestreams. Use equipment with high efficiency and low overspray. Reduce spent filter generation by covering filters with cheesecloth. Apply spent solvent waste minimization techniques to dirty thinner. Use gun washer equipment for equipment cleaning. Train spray operators to use proper application techniques.
Ignitable wastes	Use materials in concentrations below ignitability levels. Allow saturated rags and similar materials to dry before disposing. Use water-based materials and nonhazardous solvents. Segregate incompatible wastes to avoid combustion or reaction.
Vehicle maintenance wastes	See Sec. 9.13 for a discussion and detailed checklist (Table 9.12) of waste minimization alternatives that may be applied to vehicle maintenance wastes.

TABLE 9.2   Checklist of Waste Minimization Alternatives for Dry Cleaners and Laundries

Waste category	Alternatives
Operating practice	Improve operating practices. Make employees aware of the need to minimize the generation of hazardous wastes. Establish preventive maintenance programs. Implement procedures to minimize the loss of solvents. Use solvents with flash points lower than 140°F. Periodically replace seals and gaskets. Open button traps and lint baskets only briefly to avoid residual solvent losses. Use equipment with monitors to ensure application of the correct amounts of solvent. Use dry to dry machines and/or cold dry systems. Install equipment to remove water from oil and grease sludges.
Housekeeping	Implement good housekeeping procedures. Check for and fix leaks regularly. Repair holes in air and exhaust ducts. Keep containers of solvent closed while not in use. Clean lint screens regularly to avoid clogging of fans and condensers.
Recycling	Install an on-site solvent distillation unit. Ship waste solvent off site to a solvent reclaimer. Return waste solvent to supplier.

9.4   Educational and Vocational Facilities

Educational and vocational facilities generate many different kinds of hazardous wastes due to the variety of activities that take place in schools. Hazardous waste categories that can be found in educational and vocational facilities are listed below:

- Laboratory wastes
- Printing wastes
- Photographic wastes
- Vehicle maintenance wastes
- Paint wastes
- Metalworking wastes

Table 9.3 provides a checklist of waste minimization alternatives that apply to educational and vocational facilities. Additional waste minimization options may be found in other sections of this chapter which address industry groups that may be related to activities conducted at schools.

TABLE 9.3   Checklist of Waste Minimization Alternatives for Educational and Vocational Facilities

Waste category	Alternatives
Laboratory wastes	See Sec. 9.7 for a discussion and detailed checklist (Table 9.6) of waste minimization alternatives that may be applied to laboratory wastes.
Printing wastes and wastewaters	See Sec. 9.12 for a discussion and detailed checklist (Table 9.11) of waste minimization alternatives that may be applied to printing wastes.
Photographic wastes	See Sec. 9.10 for a discussion and detailed checklist (Table 9.9) of waste minimization alternatives that may be applied to photographic wastes.
Vehicle maintenance wastes	See Sec. 9.13 for a discussion and detailed checklist (Table 9.12) of waste minimization alternatives that may be applied to vehicle maintenance wastes.
Paint wastes	See Sec. 9.2 for a discussion and detailed checklist (Table 9.1) of waste minimization alternatives that may be applied to paint wastes.
Metalworking wastes	See Secs. 9.6 and 9.8 for discussions and detailed checklists (Tables 9.5 and 9.7) of waste minimization alternatives that may be applied to metalworking wastes.

9.5   Electroplating

The hazardous wastes generated from electroplating processes can be grouped to reflect four different sources:

- Work cleaning wastes
- Spent plating solutions and sludges
- Waste rinse water
- Treatment wastes

Table 9.4 presents a checklist of waste minimization alternatives for electroplating operations. Additional waste minimization options may be found in other sections of this chapter which address industry groups that may be related to activities conducted in the electroplating industry.

9.6   Fabricated Metal Manufacturing

The fabricated metal manufacturing industry produces hazardous waste in most of its operations. Examples of the hazardous wastes generated are listed below:

TABLE 9.4   Checklist of Waste Minimization Alternatives for Electroplating Operations

Waste category	Alternatives
Work cleaning wastes	Reduce cleaning frequency. Design process and equipment to minimize surface area exposed to process fluid. Record cleaning costs as a separate item. Convert from a batch process to a continuous process. Maximize dedication of process equipment. Avoid unnecessary cleanups. Operate equipment to inhibit fouling. Minimize residue build-up during operation. Minimize the amount of cleaning solution used. Recycle cleaning solution by filtering solids from used solution.
Spent plating solutions and sludges	Increase plating solution life. Substitute cyanide-free plating solutions for cyanide solutions. Replace cadmium-based solutions with zinc solutions. Use trivalent chromium in place of hexavalent chromium, when possible. Return spent plating solution to manufacturer.
Waste rinse water	Operate equipment to reduce drag-out. Increase solution temperature. Lower the concentration of plating bath constituents. Consider use of an air knife to reduce surface evaporation. Reduce speed of product withdrawal to allow for drainage time. Use surfactants to lower solution surface tension. Properly position workpiece on rack for maximum drainage. Recover drag-out of plating solution. Use multiple rinse tanks in countercurrent series. Use fog nozzles and sprays for coating simple workpieces. Reuse rinse water, when possible. When still-rinsing, recycle rinse water upstream. Use automatic flow instrumentation to control flow rate. Increase rinsing efficiency by agitating rinse bath.
Treatment wastes	Substitute different precipitating agents. Use trivalent chromium in place of hexavalent chromium. Segregate waste streams to facilitate treatment and recovery of metals. Install a sludge dewatering system. Improve operating practices. Install a metal recovery system.

- Spent solvents and solvent still bottoms
- Paint wastes with heavy metals
- Acid and alkaline wastes
- Waste oils
- Spent abrasives

- Metal dusts, grindings, and cuttings

Table 9.5 provides a checklist of proven waste minimization practices and techniques that may be used in fabricated metal manufacturing. Additional waste minimization options may be found in other sections of this chapter which address industry groups that may be related to fabricated metal manufacturing.

9.7 Laboratories

Laboratory wastes can be classified into five categories:

- Spent solvents
- Unused reagents
- Reaction products
- Test sample wastes
- Contaminated materials

Because of the diversity of characteristics exhibited by the types of hazardous wastes listed above, laboratories must use several different management methods, which is more expensive and time-consuming than handling only one waste type. Therefore, waste minimization can play an important role in a laboratory waste management strategy.

Table 9.6 provides a checklist of waste minimization alternatives for laboratories. Additional waste minimization options may be found in other sections of this chapter which address industry groups that may be related.

9.8 Metal Finishing

Metal-finishing wastes result from processes that treat and coat metal surfaces. Hazardous wastestream categories for the metal-finishing industry are listed below:

- Spent heat-treating agents
- Electroplating/anodizing/galvanizing wastes
- Chemical finishing wastes

Table 9.7 provides a checklist of waste minimization alternatives for metal-finishing operations. Additional waste minimization options are provided in the electroplating section and other sections of this chapter which address industry groups that may be related.

TABLE 9.5   Checklist of Waste Minimization Alternatives for Fabricated Metal Manufacturing Operations

Waste category	Alternatives
Spent solvents and solvent still bottoms	Install lids or silhouettes on tanks. Increase freeboard space on tanks. Install freeboard chillers on tanks. Remove sludge from solvent tanks frequently. Preclean parts manually to increase solvent life. Use a solvent recycling service. Consider using on-site solvent distillation units. Substitute less hazardous solvents and degreasers. Slowly remove parts from vapor zone. Avoid cross-contamination of solvent. Use mechanical blasting in place of solvent stripping. Reduce use of metals-based paints, when possible. Use water-based coatings. Use high solids coatings such as powder coatings. Implement better operating practices for painting. Train spray operators on proper application techniques. Segregate all wastestreams. Use equipment with high efficiency and low overspray. Reduce spent filter generation by covering filters with cheesecloth. Use gun washer equipment for equipment cleaning.
Acid and alkaline wastes	Use deionized water to prepare solutions. Remove sludge continuously. Install lids on tanks. Use water sprays and fog nozzles, where possible. Avoid spillage and dripping of acid or alkaline cleaning solutions. Segregate acid and alkaline wastes. Standardize oil types used on machining equipment. Use dedicated lines or improve scheduling of equipment use. Reuse or recycle cutting, cooling, and lubricating oils. Substitute a hot lime bath or borax soap for lubricating oils. Consider separating oils from metal fines through high-speed spinning action and filtration.
Spent abrasives	Use water-based or greaseless binders. Apply abrasives to wheel automatically to control application rate. Ensure sufficient water use during cleaning.
Metal dusts, grindings, and cuttings	Segregate scrap metal.

TABLE 9.6 Checklist of Waste Minimization Alternatives for Laboratories

Waste category	Alternatives
Spent solvents	Use on-site distillation and filtration equipment.
	Substitute nonhazardous or less hazardous solvents and cleaners.
	Consider purchasing smaller lots and quantities.
Unused reagents	Purchase only necessary quantities (inventory control).
	Substitute nonhazardous or less hazardous materials.
	Store chemicals in a well-organized, central location.
	Use older chemicals first.
	Maintain a list of all unused chemicals for waste exchange.
	Use only necessary quantities of chemicals.
	Make chemical substitutions, where possible.
	Provide separate containers for reagents and reaction products that are suitable for recycling.
	Store chemicals in a well-organized, central location.
Test sample wastes and contaminated materials	Use only necessary quantities of chemicals.
	Make chemical substitutions, where possible.
	Reduce sample sizes.
	Use proper laboratory methods and techniques.
	Check equipment cleaning procedures.

TABLE 9.7 Checklist of Waste Minimization Alternatives for the Metal-finishing Industry

Waste category	Alternatives
Heat-treating wastes	Replace barium or cyanide salt baths with alternate treatment methods.
	Use more dilute process solutions.
	Use filtration process for oil quench baths.
	Replace solvents with alkaline washes.
	Extend life of alkaline wash by removing oil.
Electroplating/ anodizing/ galvanizing wastes	See Sec. 9.5 for a detailed discussion and checklist (Table 9.4) of waste minimization alternatives that may be applied to these wastes.
	Use better operating practices.
	Extend bath life.
	Recover metals and acid from spent baths.
	Spray or brush workpiece instead of immersion.
	Use alternative treatment techniques.
	Use less toxic process solutions.
	Use more dilute process solutions.
	Reduce drag-out.
	Employ effective rinsing methods.
	Use immiscible rinses.
	Use no-rinse coatings.
	Recycle spent rinse water.
	Reclaim metals from waste rinse water.
	Change rinse composition.
	Reclaim metals from filter wastes.
	Employ effective dewatering process for filter wastes.
	Reduce production of off-specification coatings.
	Use nonchrome etchants.

9.9 Pesticide Users

Pesticide users have various opportunities to minimize the hazardous wastes they must manage. Hazardous wastes generated by pesticide users are listed below:

- Waste rinse water and absorbent
- Off-specification products
- Packages and drums
- Dust collected in air pollution control equipment

Most waste minimization practices that apply to the pesticide industry center around good operating practices. Table 9.8 provides a checklist of waste minimization alternatives for pesticide users. Additional waste minimization options may be found in other sections of this chapter which address industry groups that may be related to activities conducted by pesticide users.

9.10 Photography

Hazardous photographic wastes can be grouped into three categories:

- Process bath wastes
- Color developer wastes
- Bleach/fix/bleach-fix wastes

Most waste minimization activities that apply to the photography industry are centered around good operating practices, recycling, and recovery. Commercially available recycling equipment makes it pos-

TABLE 9.8 Checklist of Waste Minimization Alternatives for Pesticide Users

Waste category	Alternatives
Waste rinse water or absorbent	Reuse waste rinse water in future applications. Use high-pressure spray nozzles. Use dry absorbents for cleanup. Sweep floor to collect spilled product for use in reformulation.
Off-specification products	Control formulation process carefully. Reformulate off-specification products instead of disposing. Use minimal amount of rinse water. Recondition and reuse drums. Use refillable or returnable bulk tote bins.
Dust collected in air pollution equipment	Recycle dust into process from which it was generated.

TABLE 9.9   Checklist of Waste Minimization Alternatives for the Photography Industry

Waste category	Alternatives
Process bath wastes	Use good operating practices. Use metallic replacement systems to settle out silver. Use chemical precipitation products. Install metal recovery system.
Color developer wastes	Use metal recovery techniques. Use countercurrent rinsing. Use squeegees to wipe excess liquid from moving photographic material. Reduce water consumption.
Bleach/fix/bleach-fix wastes	Use ozone oxidation to regenerate spent ferricyanide bleach. Use electrolysis to regenerate spent ferricyanide bleach. Use electrolysis to desilver the bleach. Use persulfate salts to regenerate spent ferricyanide bleach. Use liquid bromine to regenerate spent ferricyanide bleach. Use iron-complexed bleaches to replace ferricyanide bleaches.

sible to reuse spent materials. Also, systems are on the market that allow for the recovery of silver from chemical solutions used in the industry.

Table 9.9 provides a checklist of waste minimization alternatives for the photography industry. Additional waste minimization options may be found in other sections of this chapter which address industry groups that may be related.

9.11   Printed Circuit Board Manufacturing

Hazardous wastes that are generated by the printed circuit board manufacturing industry result from the five operations listed below:

- Cleaning and surface preparation
- Catalyst application and electroless plating
- Pattern printing and masking
- Electroplating
- Etching

Table 9.10 provides a checklist of waste minimization alternatives for printed circuit board manufacturing operations. Additional waste minimization options are provided in other sections of this chapter which address industry groups that may be related.

TABLE 9.10   Checklist of Waste Minimization Alternatives for the Printed Circuit
Board Manufacturing Industry

Waste category	Alternatives
Cleaning and surface preparation	See Table 9.1 for waste minimization alternatives for solvents and acid and alkaline cleaners. Use water-based binders to minimize abrasive wastes. Apply abrasives to grinding wheel with automatic liquid spray. Control water levels in grinding and cutting equipment. Use good operating practices when washing and rinsing.
Catalyst application and electroless plating	Combine sensitization and activation solutions to reduce rinsing steps. Use lower concentration plating baths for easier rinsing. Use differential electroless plating. Use weak or biodegradable chelating agents. Use in-line metal recovery techniques. Use automated control systems for board handling and process bath monitoring.
Pattern printing and masking	Use water-processable resist instead of solvent-processable resist. Substitute screen printing for photolithography. Use Asher dry photoresist removal method.
Electroplating	See Sec. 9.5 for a discussion and detailed checklist (Table 9.4) of waste minimization alternatives for electroplating operations.
Etching	Use dry plasma etching techniques. Use additive instead of subtractive method. Use less toxic etchants, where possible. Install in-line metal recovery system. Use thinner layer of copper foil on laminated boards.

9.12   Printing

Printing industry hazardous wastes can be placed in three categories:

- Trash
- Wastewater
- Equipment cleaning wastes

Waste minimization activities that apply to the printing industry are centered around good operating practices, product substitution, recycling, recovery, and equipment modification. Table 9.11 provides a checklist of waste minimization alternatives for the printing industry. Additional waste minimization options may be found in other sections of this chapter which address industry groups that may be related.

TABLE 9.11   Checklist of Waste Minimization Alternatives for the Printing Industry

Waste category	Alternatives
Trash	Recycle empty containers or return them to supplier for refilling. Use proper storage procedures for photosensitive materials. Recycle spoiled photographic film and paper. Monitor press performance carefully. Install web break detectors. Use electronic imaging and laser platemaking processes.
Wastewater	Use silver-free films. Use water-developed lithographic plates. Recycle spent solutions. Recover silver from spent solutions. Employ countercurrent washing. Use squeegees to wipe off excess liquid in nonautomated processing systems. Substitute iron-EDTA for ferrecyanide bleaches. Implement better operating practices. Remove heavy metals from wastewater. Use electronic imaging and laser platemaking processes. Use washless processing systems.
Equipment cleaning wastes	Substitute less toxic or nonhazardous solvents. Schedule jobs to reduce number of cleanups. Recycle waste ink and cleanup solvent. Use automatic cleaning equipment. Use automatic ink leveler to reduce ink waste and spillage.

9.13   Vehical Maintenance

Most aspects of the vehicle maintenance industry involve generation of hazardous wastes. Wastestreams from the vehicle maintenance industry are listed below:

- Parts cleaning wastes
- Waste oils and coolants
- Spent lead-acid batteries
- Paint wastes

Table 9.12 provides a checklist of waste minimization alternatives for the vehicle maintenance industry. Additional waste minimization options may be found in other sections of this chapter which address industry groups that may be related.

TABLE 9.12   Checklist of Waste Minimization Alternatives for the Vehicle
Maintenance Industry

Waste category	Alternatives
Parts cleaning wastes	Operate solvent sinks properly. Avoid contamination of solvent. Install lids/silhouettes on solvent tanks. Increase freeboard space on tanks. Use commercial solvent recycling service. Employ on-site recycling of solvent using distillation and/or filtration. Substitute biodegradable cleaners for solvents. Remove sludge frequently from cleaning tanks. Use multiple rinse tanks. Recycle rinse solutions. See Table 9.1 for additional waste minimization alternatives for solvents and acid and alkaline cleaners.
Waste oils and coolants	Change fluids only at required intervals. Store oil separately from other hazardous wastes. Recycle used oil through a reputable recycler. Burn used oil in on-site heating units, if allowable. Extend oil change intervals with fluid filtration systems. Avoid spillage. Use drip pans to catch leaking fluids. Filter particulates from antifreeze and reuse, where possible.
Spent lead-acid batteries	Return to manufacturer or recycling facility for reconditioning.
Paint wastes	Reduce use of solvent and metals-based paints, where possible. Use water-based coatings. Use high-solids coatings such as powder coatings. Implement better operating practices for painting. Segregate all wastestreams. Use equipment with high efficiency and low overspray. Reduce spent filter generation by covering filters with cheesecloth. Apply spent solvent waste minimization techniques to dirty thinner (see Table 9.1). Use gun washer for equipment cleaning. Train spray operators on proper application techniques. Use materials in concentrations below ignitability levels. Allow saturated rags and similar materials to dry before disposing. Use water-based materials and nonhazardous solvents. Segregate incompatible wastes to avoid combustion or reaction.

9.14 BIBLIOGRAPHY

American Chemical Society: *Less Is Better: Laboratory Chemical Management for Waste Reduction*, Washington, D.C., 1985.

Center for Hazardous Materials Research: *Waste Minimization Manual for Pennsylvania's Vehicle Maintenance Industry*, Pittsburgh, Pa., 1987.

Center for Hazardous Materials Research: *Waste Minimization Manual for Small Quantity Generators*, Pittsburgh, Pa., 1987.

Erie County Department of Environment and Planning: *Guide for Treating, Storing, Shipping, and Disposing Hazardous Wastes from Educational and Vocational Schools*, Buffalo, N.Y., 1987.

Marketing
Waste Reduction

David Wigglesworth

Program Manager,
Waste Reduction Assistance Program (WRAP)
Alaska Health Project

Profit is a meaningless statement of the
corporate purpose. Without customers in
sufficient and steady numbers there is no
business and no profit. No business can
function effectively without a clear view of
how to get customers, what its prospective
customers want and need, and what options
competitors give them, and without explicit
strategies and programs focused on what
goes on in the market place rather than on
what's possible in the factory or what is
merely assumed at headquarters.

THEODORE LEVITT
The Marketing Imagination

Today, a trash dumpster is easier to locate than a recycling bin in most communities across the country. This situation reflects a society which regards wastes of all types as things to throw away. Certainly this is not a new concept, but it is a fundamental issue to bear in mind when discussing the integration of waste reduction into the fabric of our industrial society.

Waste reduction can be defined as any activity that reduces the amount of waste generated at its source in the production process— before the waste becomes a pollutant or a "lost" resource. Acceptance of waste reduction will require a dramatic change in attitude among individuals and society to enable us to view wastes as valuable resources. It will require the creation of an infrastructure within our institutions, both public and private, which promotes waste reduction to the fullest extent. It will require the transfer of existing technology and technological innovation. It will require shifts in funding priorities as well.

These changes, while necessary, are not sufficient to achieve nationwide acceptance of waste reduction and continued use of waste reduction techniques. Equally important if not more important in consideration of this goal are the methods used to achieve and maintain these changes. How can we overcome the "if it ain't broke, don't fix it" attitude and other barriers that prevent rapid acceptance and use of waste reduction technologies? A closer look at the topic of marketing may provide an answer.

This chapter explores the issue of marketing waste reduction as a vital component to successful integration of waste reduction into our society. It is by no means a complete analysis of the issue. It suggests various methods to market waste reduction with the hope of spurring the imagination of others in this area and promoting development of waste reduction programs.

This chapter is written primarily from the point of view of a state waste reduction delivery program. Specific examples of internal business delivery programs are provided, when appropriate, for the purposes of comparison and clarification of the information presented.

Readers should also note that common marketing terms such as customer, product, seller, and competition have been specifically defined to demonstrate the points made in this chapter and to reinforce the notion that waste reduction is as much an issue of marketing as it is an issue of technology. These marketing terms are defined as follows:

Seller   means a waste reduction delivery program, such as a state or local government program.

Customer   means a business manager, employee, community, industrial sector, or other appropriate target audience for the seller.

Product   means waste reduction.

Competition   means the obstacles or barriers to waste reduction. Examples of waste reduction competition are provided in the chapter.

Inspiration for this chapter is attributed to Theodore Levitt's book *The Marketing Imagination*, an eloquent discussion on the topic of marketing. This no-nonsense, superbly written book is well worth reading and may be helpful for others considering marketing waste reduction. Special thanks also goes to Adrian Dube for her helpful advice.

10.1   What Is Marketing?

Marketing is many things to many people. To some it is an ugly word associated with obtrusive advertising, subliminal manipulation, and a salesperson knocking at the front door just as you sit down to a great dinner. To others, it is a fascinating investigation into the manner in which people think and act.

To a business person, marketing is a vital component to an overall business plan. A business's marketing plan focuses on how the business promotes its goods and services over those of its competition. It requires that a business know who its customers are, how they think, and how its product or service will meet customer needs—ultimately making their life easier. In short, as considered by Levitt in *The Marketing Imagination*, marketing is the link between the seller and the customer which focuses on capturing and maintaining a customer to ensure the life of the service or product and a successful business venture.

10.2   What Is Marketing Waste Reduction?

Marketing waste reduction is in many respects identical to simple business marketing. The goal in marketing waste reduction, just as with a business marketing plan, is to initially attract customers and maintain their interest over time.

The chore of any seller (state program) marketing waste reduction (the product) is (1) to attract customers (industry, business managers) by making waste reduction credible in their eyes and (2) to keep customers interested in waste reduction by making it more attractive than pollution control technologies (competition). In the case of waste reduction, the competition is any obstacle or barrier to waste reduction commonly mentioned in other literature concerning waste reduction. Examples include

- Federal, state, and local environmental programs historically fixated on the use of pollution control strategies, such as treatment and land disposal
- Lack of commitment on the part of business managers and society to reduce the generation of waste, wherever and whenever possible
- Lack of funding to support development of meaningful waste reduction programs at the federal, state, and local level
- Limited number of incentives such as revolving loan programs and challenge grants that encourage waste reduction among the business community
- Business concerns about product quality when considering waste reduction technology
- Lack of guidance and technical assistance for businesses interested in identifying waste reduction opportunities in their shops
- Lack of awareness among businesses and state waste reduction programs about the availability of waste reduction technologies and the effectiveness of these technologies, or lack thereof
- Unwillingness among industry to seek assistance due to concerns about confidentiality and trade secrets
- Continued debate over the definition of waste reduction
- Lack of waste reduction curricula in secondary and postsecondary schools
- Emphasis on hazardous waste reduction versus focusing on all waste that enters the land, air, and water.

To overcome waste reduction competition, emphasis must be centered on economic, educational, regulatory, and corporate obstacles, rather than solely on the barrier of technology itself. In order to promote waste reduction, a waste reduction program must focus primarily on appealing to customer needs rather than waste reduction itself. The key to any program is defining what it is that waste reduction will offer a target audience and developing the specific programs that will make the audience choose this concept over the competition.

Once a customer has bought into the idea of waste reduction, the next critical point is to develop methods to ensure that that customer continues to buy into it, which may require different selling strategies than those that influenced the customer to accept waste reduction initially. For example, in the past, businesses bought into pollution control strategies to avoid regulatory enforcement. Today, some businesses buy into waste reduction to avoid regulatory enforcement, but also because they realize it can improve business efficiency, decrease operating costs, legitimately improve their public image, and so on. In short, the benefit package derived from buying waste reduction is superior to that derived from pollution control. The goal of any waste reduction delivery program is to ensure that this is the case—hence the need for and importance of marketing waste reduction.

10.3 Marketing Waste Reduction

While the need for marketing waste reduction is clear, exact methods on how to do it may vary depending on the target audience(s) of a waste reduction delivery program, local conditions, and other factors. The remainder of this chapter addresses key market questions that need to be considered by a waste reduction delivery program in order to build and maintain long-term support and interest in waste reduction. In summary, these questions include

1. Who is the program being sold to (i.e., who is the target audience?)?
2. How is the program packaged?
3. How does the program maintain customer interest in waste reduction?
4. How does the program encourage customers to market waste reduction on their own (in order to increase overall support for waste reduction)?

Specific examples demonstrating these points and resources for more information are also provided.

10.3.1 Who is the program being sold to?

Answering this question is critical and will influence the direction of a waste reduction delivery program and related marketing efforts. The response to this question is also dependent on a number of issues, most immediately staffing and funding levels for a program and ultimately the amount of time that program staff can spend on cultivating program support and interest. It is also influenced by the initial starting point for a program, that is, whether a delivery program is initiated within a state or local regulatory agency, a university, a nonprofit group, a community organization, or simply an internal company-specific program.

A waste reduction program can have a variety of target audiences, both long- and short-term, depending on program and customer needs. Selected target audiences for state programs include

- Large businesses
- Small businesses
- Homeowners
- Educational institutions
- Local, state, and federal government
- Business associations
- Local, state, and federal politicians
- Community organizations
- Environmental and public interest groups
- Health professionals
- Indian communities

Selected target audiences for internal business programs can include

- All of the above, for the main purpose of building awareness about the business program
- Production engineers
- Support staff
- Safety managers
- Customers
- Supply companies

Given the range of target audiences to market, it may be worthwhile initially to select an audience based upon primary need. A business program might first target production managers and engineers. A state or local program would initially focus on specific industries. It is also worthwhile to select audiences on the basis of how they can ultimately further the goal of attracting more waste reduction customers and maintaining interest over time. In this case, state small business development centers, business associations, and even schools are likely candidates. The issue of waste reduction can easily fit into the organization's own programs, simultaneously promoting its goals and those of a state program.

10.3.2 How is the program packaged?

Once target audiences have been determined, the next important step in marketing waste reduction is to convince customers to "buy" it.

Proper packaging of a delivery program is critical and must consider both the technical and behavioral obstacles to waste reduction.

Several important packaging issues to consider when developing or reviewing an existing delivery program are discussed below. These points should be considered, regardless of program location (i.e., state agency, university, etc.). These points include program location, program services, program promotion, program delivery, and program service costs.

Program location. It is clear that certain target groups will more readily identify with certain locations for waste reduction delivery programs. Fundamental in terms of locating a program is determining the needs of the selected target audience(s). For example, business may shy away from using a government-based program due to concerns about confidentiality and regulatory enforcement. These concerns become less important when delivery programs are located outside of government facilities in a university or nonprofit setting.

Availability for waste reduction funds is increasing for state agencies, however, underscoring the issue of program longevity. Consistent funding from year to year is an important criteria when selecting the proper location for a program, particularly when trying to maintain interest in waste reduction over time.

Perhaps the best solution is to establish waste reduction delivery programs at many levels and in a variety of locations. Since acceptance of waste reduction requires social change, many different individuals and organizations will be and will have to be involved to help reorient society to view wastes as resources. Thus a multilevel statewide waste reduction delivery program is important in order to:

- Heighten waste reduction visibility

- Attract the greatest number of buyers by appealing to behavioral and institutional concerns

- Increase the probability of maintaining waste reduction services over time

- Provide greater direct assistance to customers (the more people a delivery program services, the greater the impact)

- Increase support for long-term funding for waste reduction programs at all levels

- Enhance private–public partnerships for delivery of waste reduction services

In short, the creation of waste reduction service delivery programs at all levels increases knowledge about the superiority of waste reduction programs over pollution control strategies. This will ultimately help to build (market) nationwide acceptance for waste reduction.

Program services.   Program services are an important part of program packaging. Services provided should be based upon the needs of the selected target audiences and on what other waste reduction services are readily available to them.

In order to appeal to the widest range of potential customers, a corresponding range of program services is needed. Each individual service must be flexible enough to adapt to the needs of all customers. For example, fact sheets on waste reduction for business groups will need to include different information than will similar materials designed for student groups or community organizations. Moreover, the design and language of the fact sheets may have to be flexible in order to communicate effectively to varied interests within a single customer group or target audience. This is also true for newsletters, workshops, telephone hotlines, and so on. Core services offered by state waste reduction programs include

- Telephone hotlines which provide quick and accurate information (toll-free telephone numbers are useful).
- Newsletters to provide more in-depth information.
- On-site audits to provide direct assistance to business and to heighten program visibility.
- Revolving loans, tax incentives, and challenge grants to create positive incentives for use of waste reduction technology.
- Seminars and workshops to support needed continuing education.
- Library services and bibliographies to support waste reduction research.
- Pollution prevention awards to applaud public and private efforts to utilize and promote waste reduction technology.
- Fact sheets, handbooks, and other technical information.
- Waste reduction student intern programs coordinated with local engineering schools to educate future engineers about waste reduction and to increase program staff of delivery programs.
- Education outreach efforts, which typically are designed to advertise a service delivery program and encourage business and other groups to identify and utilize waste reduction options. Outreach program should also consider helping target groups market waste reduction to their own clientele. These efforts should be keyed into the state delivery program.

A company waste reduction program could utilize many of the services previously mentioned, redesigning them specifically for company employees. Other company programs to encourage overall support for waste reduction, both inside and outside of the company, could include

- Establishment of employee funds supported by a certain percentage of waste reduction savings. These funds could be split among employees, used for employee training, or used for employee social functions.

- Creation of company matching donation programs to support and encourage employee contributions to nonprofit waste reduction programs.

- Initiation of "Big Brother" programs in which one business can help another business identify waste reduction opportunities.

- Opportunities for waste reduction education and training opportunities for employees.

Program promotion. A variety of waste reduction services will certainly attract and maintain customers; however, promotional activities are also needed to increase the visibility of these services. These activities must also be packaged to meet the needs and behavioral patterns of each target audience and the range of individuals within a specific target group. For example, promotional programs for students need to be different from programs designed for small business. Student promotional programs also have to be developed according to the grade level and type of student.

Program delivery. Program delivery is also an important consideration when deciding how to package a waste reduction delivery program. Effective delivery strategies take into account the attitudinal and technical obstacles to waste reduction within a target group. This underscores the need to have a wide variety of personnel, with varied backgrounds, involved in a waste reduction service delivery program.

For example, apprehension on the part of small business owners to request an on-site waste reduction audit generally involves behavioral rather than technical barriers. Client concerns about confidentiality and enforcement must be addressed. A person with training in human relations may be better at addressing these barriers than someone with a technical background only. Thus, consideration should be given to creating an audit team that can be involved at different stages in the audit process to address client-specific behavioral and technical needs.

The same consideration applies to an internal business audit team. Frontline workers, top management, personnel and purchasing departments all have roles to play in the audit process and its results. If recommendations are made to alter plant operations to promote waste reduction, training may be necessary to support these changes. Consideration should be given to the type of person designated to conduct the training and to address employee concerns—both technical and behavioral.

Education is another key packaging component in the delivery (and promotion) of waste reduction services. An emphasis on education and technical assistance throughout program delivery to the target audience offers something that pollution control strategies typically have not provided—*help*. Perhaps the greatest fault in the delivery of pollution control strategies, aside from shifting pollution from one media to another, is the failure to deliver meaningful education programs when new programs, and new regulations in particular, are developed.

Businesses, particularly smaller ones, often do not have the capability to address regulatory changes, due to lack of funds, training, and staff. Thus, historical efforts to sell pollution control strategies to business have sometimes failed in part because the strategies did not address the educational needs and the situations of the target audience. Emphasis on education throughout a service delivery program increases the waste reduction benefit package, making it more appealing and acceptable to the customer.

Program service costs.   The costs for services is an important consideration when marketing waste reduction. The costs are part of an overall program package, particularly in the early stages of program development and delivery. As with other considerations, service fees must take into account the nature of the target audience(s). Moreover, consideration should also be given to the customers' willingness and ability to pay. Successful waste reduction programs generally provide free services for their target groups, charging only nominal fees for waste reduction seminars and publications.

10.3.3 How does the program maintain customer interest in waste reduction?

Customers generally buy a product if the price is right and the product is good. However, product reliability and price are not always enough to maintain long-term customer interest. As mentioned before, a customer buys a product and the benefit package that comes along with it. In the case of waste reduction, the program package (program activities, promotional programs, pricing, and delivery) is extremely important in order to overcome obstacles to waste reduction and to generate long-term interest in waste reduction over pollution control. Some key points for program delivery personnel to consider when attempting to develop long-term support for waste reduction are discussed below.

Consistent information. Your target audience will know what to expect from you.

Rapid response. Getting the runaround is frustrating for anybody. Customers want information as soon as possible, and every effort should be made to accommodate their needs.

Program flexibility. Being able to adapt and respond to changing customer needs and new clientele is critical.

Customer communication. Don't wait for a business to contact you about waste reduction. Delivery programs must constantly communicate with customers to maintain interest over time. Forms of communication can be both passive and active, including

Telephone calls

Radio public service announcements

Utility "bill stuffers"

Articles in newsletters and newspapers

Technical bulletins

On-site visitations

Attendance at meetings held by customers, including trade shows, business association meetings, and Chamber of Commerce programs

Regular seminars and workshops

The communication content can vary from very specific to simply informational. The key is to maintain visibility and demonstrate to customers that you are available to help and you appreciate their support.

Training the trainer. The more people trained to provide services and to promote the issue of waste reduction, the more readily it will become accepted.

Waste reduction curricula. It is important to instruct future workers and managers about the issue of waste reduction. Waste reduction curricula should continue to be developed and integrated into all levels of our educational system. In addition, waste reduction curricula can be included in a variety of other types of educational programs, including pre-business seminars offered by small business development centers, union apprenticeship training textbooks, training programs for environmental control officers, and programs for investors and inventors, to name just a few.

Client confidentiality. Maintaining the trust of your target audience is extremely important. It is also important that the target audience, for example, a business, knows right from the start of any situations in which you may have to in any way breach this agreement (for example, in situations where you discover a direct threat to human health or to the environment).

Responding to all concerns. Issues concerning waste reduction may uncover a number of other problems for target audiences. For example, when conducting a waste reduction audit, the audit team may uncover safety and health problems, record-keeping difficulties, and so forth. A waste reduction delivery program should anticipate these and other potential concerns and develop plans to address them. The important point is to help customers with their problems and not to turn them away.

Responding to all concerns may involve developing in-house capabilities or developing working relationships with other groups that can provide additional information. In the latter case, it would be important to ascertain the track record of these other groups to be certain that they are credible in the eyes of program customers.

Groups that have developed relationships with each other in the delivery of waste reduction services include

State agencies

Universities

Nonprofit groups

Environmental health organizations

Business associations

Trade associations

Professional groups

Small business development centers

Maintaining customer expectations. The more programs that are offered in the area of waste reduction, the greater will be customer expectations to have them continued (assuming that the programs are worthwhile!). This requires that a waste reduction program have a consistent budget to maintain core programs. Moreover, it requires that programs continue to change as new activities and programs are needed by customers.

Providing support. Integrating waste reduction into our society will take time, funding, and education. The starting point for accepting waste reduction will vary from audience to audience and among individuals who make up a target group. This point should be considered when marketing waste reduction and when recognizing businesses and groups for their efforts, such as through pollution prevention awards.

Even simple steps toward the goal of waste reduction should be noted, such as a business opening its doors to an outside audit team or setting up an in-house audit team. A small electroplating shop which installs a spray bar to reduce drag-out may be just as deserving of recognition as is a major company that reduces waste generation by 40 percent and realizes an annual savings of a million dollars. The small company's action may even be the more momentous event.

The important point, particularly when cultivating new target groups, is not necessarily how advanced a group's waste reduction efforts are, but that the group is doing something toward waste reduction. The goal of a service delivery program and related marketing programs is to convince groups to continue their efforts and do more.

10.3.4 How does the program encourage customers to market waste reduction on their own?

The more people marketing and actively doing waste reduction, the greater the likelihood for its acceptance and funding for state, local, and federal delivery programs. Once a waste reduction delivery program has successfully marketed waste reduction to its customers, the next step is to get those customers to market waste reduction to their own clientele. This is an important component in any waste reduction marketing program, because it helps develop and maintain the institutional support for such programs in all locations.

Success in encouraging customers to market waste reduction to their clientele is largely dependent upon how successful a delivery program is in maintaining long-term customer interest in eliminating waste. Being attentive to customers, providing answers to their questions, and being supportive of their waste reduction efforts is important. Perhaps most significant is that a delivery program have a true understanding of its customers; the program must know how waste reduction can fit into the customers' activities and actively develop creative approaches for customer consideration.

For example, universities should be encouraged to develop waste reduction curricula and provide waste reduction internship opportunities for graduate engineering students. Students benefit from the opportunity to become more directly involved in this area and do some hands-on work. Universities may be able to increase the number of students in their programs by offering this type of program, along with a small stipend and credit for student work. In short, students benefit from the educational opportunity, a university benefits by attracting more students to its programs, and the overall goal of promoting waste reduction is benefited by university and student activity in this area.

As another example, organized labor could be approached to promote waste reduction through contract negotiations. Techniques such as product substitution and process changes can eliminate waste generation and lead to improved safety and health conditions for workers. This fact is particularly appealing to organized labor, and also to management.

Finally, businesses with active waste reduction programs should be encouraged to advertise this fact to their clienteles. There is strong nationwide support for preventing environmental pollution. By advertising efforts to reduce pollution, a business might appeal to consumer concerns in this area and attract more clients giving it an extra edge in a competitive market. This is particularly true for small businesses, such as dry cleaners and print shops. Moreover, this type of advertising also helps to educate the general public about waste reduction opportunities, increasing societal expectations about pollution prevention and possibly forcing other businesses to develop programs to reduce waste in order to maintain their clients.

Delivery program customers may need help with developing marketing strategies for their various clienteles. Provided below is one approach to assisting customers in this regard. This approach is only a suggested format to spur other ideas in this regard.

Step I: Profile your target audiences (customers).
- a. Collect literature on target audience activities, mission statements, major events, causes, philosophies, etc.
- b. Select audiences to focus on.

Step 2: Review collected literature.
  a. Determine how waste reduction can fit into the profile of selected target audience.
  b. Determine who target audience's clientele is.
  c. Identify what target audience needs to enable it to market waste reduction to its clientele.

Step 3: Package information collected in steps 1 and 2.
  a. Develop specific marketing packets and ideas that the target audience can use to market waste reduction.
  b. Analyze these materials for readability, applicability, and appeal to the target audience and its clientele.
  c. Identify key individuals within the target group to work with.
  d. Distribute packets to key individuals in the target group for comments.

Step 4: Discuss options with target audience.
  a. Meet with key individuals to discuss packets and get their comments.
  b. Generate their support.
  c. Set up meeting with entire target group.

Step 5: Meet with entire target audience
  a. Have key individual lead the meeting to encourage audience acceptance of the concept.
  b. Provide time for group discussion and input.
  c. Revise packet and hold another meeting, if necessary.

Step 6: Institutionalize waste reduction marketing program in target audience.
  a. Become active in their activities.
  b. Communicate regularly with the group.
  c. Recognize their efforts in promoting waste reduction.
  d. Market their programs along with your own, if appropriate.

10.4 Marketing Techniques to Overcome Selected Obstacles to Waste Reduction

The following case studies are designed to illustrate some of the points made previously in this chapter. The case studies are written from the point of view of a state waste reduction delivery program. They are organized into three parts: (1) the issue, (2) obstacles to waste reduction, and (3) a marketing solution. These examples are educational only and do not provide a complete analysis of all the issues that may be outlined or implied. Moreover, there may be many solutions to the problems addressed in the case studies other than those presented.

10.4.1 Case Study 1: The Small Business Audit

The issue. A small business owner is apprehensive about letting an audit team into his or her shop.

Selected obstacles

- The owner fears enforcement action from the audit team.
- The owner does not understand waste reduction.
- The owner's business is busy, and the owner sees the audit as just "one more thing to do."
- The owner feels that the business is operating at optimal efficiency.

Marketing solution. Nontechnical concerns are often the major obstacles for any audit team attempting to conduct a waste reduction audit of a small business, particularly if the audit team is soliciting the audit itself. At this point during the audit, it would be useful if the audit team included someone with expertise in human relations and communication. To overcome the obstacles in this case study a trusting relationship needs to be established between the audit team and the business owner *before* an audit can occur. Establishing this trust may take time and will most certainly take energy. However, by establishing a trusting relationship with the owner the audit team is ultimately building support for waste reduction over pollution control.

Typically, as trust is established, a small business owner becomes relieved that the audit team is there to help make the business more efficient and to provide constructive criticism without regulatory enforcement. Owners also appreciate the opportunity to speak directly with someone about these issues. A small business owner needs the opportunity to be educated and to voice concerns about waste management issues, just as the rest of the public does. The audit provides a vehicle for small business owners to fulfill part of this need. The on-site audit can be a very effective waste reduction marketing tool.

Some specific ideas for overcoming some of the obstacles in this case study are as follows:

- Locate a state waste reduction delivery program in a nonregulatory setting, such as a university, a nonprofit organization, or a nonregulatory state agency. This will help overcome any initial enforcement/compliance fears. Regulations and enforcement programs are important, but it is best to separate them from waste reduction programs.

- Spend time with the business to talk about the audit and the owner's concerns and to educate the owner about the benefits of waste reduction.

- Get a letter of confidentiality from the regulatory or granting agency. Often waste reduction delivery programs are funded by regulatory agencies. A letter from the regulatory agency stating that audit reports are not required to contain any specific information on the identity of the business can help satisfy a business owner.

- Use other company owners as references to help legitimize the audit team in the eyes of the owner of the business to be audited.

- Allow the business owner to have the first review of any audit report to ensure confidentiality and accuracy from the owner's point of view.

- Establish an audit timeline that fits into the business owner's schedule. (It is important not to rush the owner into this.) Scheduling to suit the owner will allow the owner to be attentive to the audit team.

- Provide ongoing assistance once the audit has been completed. The business may need help implementing waste reduction options identified by the audit team and securing loans or grants available to help finance report recommendations.

10.4.2 Case Study 2: Waste Reduction Curriculum in Secondary Schools

The issue. It is difficult to get teachers to use a waste reduction curriculum.

The obstacles

- Teachers are unfamiliar with waste reduction and the curriculum.
- Teachers are constantly "bombarded" with new curricula.
- There is a lack of instructor training to teach the curriculum.
- There is a lack of school administration support for the curriculum.

- Students do not see the relevance of the information.

Marketing solutions.   Integration of waste reduction curriculum into all levels of education is needed to encourage greater use and support for waste reduction. Exposing high school students to the topic is important, because many graduates enter the work force instead of going to college. Industrial arts and distributive education programs are good avenues for introducing this type of curriculum into the high school setting.

Specific techniques to achieve acceptance of waste reduction curriculum in the high school setting include the following:

- Develop a curriculum specific to the needs of your state and teachers.

- Include teachers and students in the development of the curriculum.

- Ensure that the curriculum includes sample lesson plans and plenty of examples that teachers can use in class.

- Include a resource library and audiovisual aides with the curriculum. Key this information into each lesson plan.

- Provide ongoing assistance to instructors in the event that they need help with any of the lessons or concepts. Offer to help arrange for guest lecturers to help teachers instruct the class.

- Pilot-test the curriculum prior to its formal implementation. A waste reduction delivery program should pilot-test the program to demonstrate its use and to work out any of the "bugs." Revise the curriculum, as necessary.

- Develop a teacher training course to instruct teachers and other school officials in the proper use and purpose of the curriculum. Eight-hour in-service training could provide a vehicle for training. However, a one-credit (15-hour) graduate-level course provides a better opportunity for training. First, it provides more time for training teachers. Often in-service days are filled with other opportunities for teachers. Second, the graduate-level course can apply toward teacher recertification credit requirements. Third, the graduate class typically requires an out-of-class assignment. In this case the assignment would be for teachers to replicate all or part of the curriculum in their own classrooms. This will help overcome some of the obstacles to its use and provide additional program evaluation. Moreover, school maintenance staff, administrators, nurses, and even Department of Education personnel should be encouraged to participate in the credit class. Information contained in a waste reduction curricula extends well beyond the walls of a classroom.

This same information can be used by school administrators and staff to address schoolwide waste management issues. Department of Education personnel can use this information during their review of school curricula and periodic facility evaluations. A school, similar to any business, generates waste and can benefit from waste reduction.

- Maintain efforts to train and retrain teachers to use the curriculum and to secure statewide use of the curriculum.

10.5 Resources for Further Assistance

Techniques to market waste reduction are evolving. Many individuals and organizations, private and public, have good ideas to encourage nationwide support for waste reduction. Readers are encouraged to contact state, federal, and local government waste reduction programs for additional information. For starters, readers might initially contact:

- EPA Pollution Prevention Office, Office of Policy, Planning, and Evaluation, Environmental Protection Agency, 401 M Street SW, Washington, D.C. 20460.

- The Local Government Commission, 909 12th Street, Suite 203, Sacramento, Ca. 95814, (916) 448-1198.

- The National Roundtable for State Waste Reduction Programs, 512 N. Salisbury Street, P.O. Box 27687, Raleigh, N.C. 27611, (919) 733-7015.

- Alaska Health Project, Waste Reduction Assistance Program (WRAP), 431 West Seventh Avenue, Suite 101, Anchorage, Alaska 99501, (907) 276-2864.

Finally, readers are also encouraged to read books on marketing in order to develop additional ideas for marketing waste reduction. *The Marketing Imagination*, by Theodore Levitt, is highly recommended. It is available through The Free Press, Division of Macmillan, Inc., 866 Third Avenue, New York, New York 10022, or through any bookstore.

Waste Minimization: Public Sector Activities

Reducing Hazardous Wastes: Strategies and Tools for Local Governments

Susan Sherry

Senior Policy Consultant,
California Local Government Commission

11.1 Local Government Involvement in Hazardous Waste Minimization

In response to community concerns, a growing number of local governments have begun to regulate the use, handling, release, and disposal of industrial chemicals within their jurisdictions. The regulatory control over toxics, once the sole domain of the states and federal government, is fast becoming a partnership among all three levels of government as citizens and local officials alike press for more direct involvement in solving toxics problems. For the most part, local government initiatives have developed in response to a crisis, for example when chemical contamination threatened a public water supply, emergency response personnel needed more information to fight chemical fires, or storage practices contaminated local soils. In some regions, state governments, recognizing that local agencies can respond most quickly to problems, have delegated special authority to their cities, counties, and local public districts.

Nowhere are these trends more evident than in California. Every mid-to-large-size jurisdiction in the state runs some type of major toxics control program—from underground storage surveillance and emergency response planning to hazardous waste generator inspections and industrial wastewater programs. Yet, as California local officials become more adept at solving toxics problems, they are also asking some tough questions. With limited resources, how can local communities continually respond to problem after problem? Are there preventive measures local governments can take? What are the limits of controlling and containing pollutants after they are produced?

It was in this context that California local governments first began to examine their role in industrial hazardous waste minimization. The idea was given an additional boost as local officials came to realize that minimization programs could reduce the need to find sites for large and unpopular hazardous waste management facilities. Ventura County was the first California jurisdiction to translate these insights into action. Through a two-year effort, Ventura County reduced the volume of hazardous wastes shipped off site by 70 percent, providing a much-needed success story for the rest of the state and nation.

Since 1986, over 60 other California local governments have adopted formal policy resolutions encouraging area industries to practice hazardous waste minimization. Responding to this growing interest, the Local Government Commission (LGC), a statewide organization of local elected officials, began to help localities turn this initial enthusiasm into operating programs. Described below, the model programs and policies promoted by the LGC can be used by concerned local governments in any region of the nation.

11.2   Three Program Models

Local governments, including cities, counties, sewer and water agencies, air pollution control agencies, and other special districts, can develop a wide range of programs to encourage hazardous waste minimization. With the diversity of options available, a local agency of any size or means can institute some type of an effort. These programs can be grouped into three broad categories:

- Educational programs
- Technical assistance programs
- Regulatory programs

The most effective way for a local government to help area industries reduce waste is to design an overall effort that combines selected features from all three of the above program models. Unfortunately, however, many local governments and special districts do not have the resources, technical expertise, or political support to develop a comprehensive program, at least not right at the start. If a comprehensive program is currently out of reach, local governments should still evaluate the opportunities presented by each program model. That exercise will indicate where to start, what is doable, and where an initial success might eventually lead. In short, local governments should start where they can, and build.

11.2.1 Educational programs

Educational programs use information and outreach to encourage firms to reduce their waste voluntarily. These programs employ all the traditional tools that a local government has at its disposal to communicate with the business community and general public. The best educational programs offer financial and technical information and also provide opportunities for creative and independent action on the part of industry. Educational programs are less costly than technical assistance and regulatory approaches and, for that reason, may be the most appropriate starting point for many local governments.

Table 11.1 provides a list of 28 ways for local governments to promote hazardous waste minimization through education and outreach. The options are ordered from less costly to more costly, although none requires significant expenditures. Conducting workshops, handing out waste minimization literature as part of local inspections, promoting hazardous waste minimization as part of the permitting process, establishing a blue-ribbon task force, and developing a waste minimization library are just a few of the suggested activities.

An effective educational program can be implemented on a very modest budget. To keep costs down, local officials need to design programs with two points in mind:

1. *The activities need to be incorporated into the normal, day-to-day routines of the local government:* In other words, the program

TABLE 11.1 Twenty-Eight Ways to Promote Hazardous Waste Minimization at the Local Level Through Education and Outreach*

1. Adopt a resolution supporting hazardous waste minimization.
2. Establish an internal (staff) task force.
3. Adopt an implementing resolution establishing an educational program.
4. Encourage local business leaders to set up programs.
5. Encourage other outside organizations to participate.
6. Make a specific, written request of individual companies.
7. Recruit volunteer(s) to staff effort.
8. Compile and promote industry success stories.
9. Publicly recognize achievements.
10. Work with engineering firms.
11. Organize an in-plant tour.
12. Work with local media.
13. Formally declare a hazardous waste minimization week.
14. Host a waste minimization exhibition fair.
15. Establish a blue ribbon task force on waste minimization.
16. Provide public education on consumer waste reduction.
17. Develop a waste minimization library.
18. Conduct a series of workshops.
19. Compile and distribute information.
20. Organize a speakers bureau.
21. Disseminate a newsletter.
22. Develop a local slide show or video.
23. Encourage financial assistance for waste minimization.
24. Create economic incentives.
25. Spur development of cooperative ventures.
26. Map out program to reduce waste generated internally.
27. Include waste minimization education as part of local inspections.
28. Develop plan for program expansion.

*A local government handbook entitled *Low Cost Ways to Promote Hazardous Waste Minimization* describes each of these 28 low-cost activities and provides a complete list of available resources to help with implementation. See Figure 11.2 (page 218) for the reference citing.

should be able to be implemented by existing staff people in their current roles. None of the activities should require special staff training or expertise.

2. *Outside entities, particularly business organizations and universities, should be encouraged to sponsor as many of the program activities as possible:* A local government can serve as a catalyst and coordinator. This does not mean that it has to staff every piece of the effort. Artfully delegating responsibilities to willing partners extends the limited resources available for the program and brings other institutions into the leadership network.

11.2.2 Technical assistance programs

Through assistance provided on a one-on-one basis or in small groups, these local government programs help firms identify and evaluate their site-specific opportunities for hazardous waste minimization. Since these efforts require trained staff who can provide specific advice, only those local agencies with an existing technical capability can run this type of a program in a cost-effective way. For example, most local wastewater treatment plants, air pollution control agencies, and environmental health departments have staff who are fairly knowledgeable about industrial processes. This is not the case for other local agencies, such as fire departments and planning departments.

Due to training costs and the extra staff time needed to provide the assistance, this option requires more financial support than the educational program described above. However, technical assistance programs are likely to yield more significant and lasting results because of the individual attention given to area firms. Industry participation in technical assistance programs is voluntary.

Local governments typically offer this assistance (1) in a small workshop setting with several similar firms in attendance; (2) as part of a firm's application process for a local permit, permit renewal, or business license; (3) as part of a compliance conference or enforcement proceeding; or (4) at the site of a firm's plant or shop. Local governments have found that assistance given prior to the modification or construction of a building or process line has the potential for significantly reducing hazardous waste.

On-site advice has been particularly popular with some local governments. These sessions can be either separated from or integrated into a local agency's routine enforcement inspections. Assistance incorporated into inspections is likely to be less costly but also less in-depth than advice given in separate sessions. Some localities have considered combining the two approaches, with field inspectors providing an introductory level of assistance and specialized staff working on more complex problems outside of the inspection process. In other areas, local officials have purposely separated consultation sessions from inspections so that technical assistance and regulatory activities are not the responsibility of the same staff or department.

Local governments should understand that the assistance provided does not need to be sophisticated. In most cases, local government staff would not attempt to provide technical solutions to complex industrial processes. Rather, they would ask probing questions, point firms in the right direction, and suggest organizational changes that would promote waste minimization. Direct advice would be given only on basic practices and technologies. As staff become more experienced, a higher level of advice can be given.

Even though most local governments will choose to emphasize "low-tech" assistance, especially at the beginning of a program, their staffs will still need some additional training. Table 11.2 lists six cost-effective ways to provide staff education in waste minimization.

Some communities have begun to explore their role in waste minimization financial assistance activities. Through its own staff, a non-profit corporation, a business–government partnership, or some other entity, a local government can try to actively help small and midsize businesses secure long-term, low-interest loans and other financial assistance for waste minimization gains.

11.2.3 Regulatory programs

Local governments can promote waste minimization through the use of existing regulations or through the development of new requirements. Codes, ordinances, licenses, and permits are a few of the tools that local governments can use for this purpose. Discussed below, these local regulatory options fall into one of three categories:

- Direct regulatory requirements
- Indirect regulatory inducements
- Positive incentives

The regulatory options are most effective when required in conjunction with a technical assistance program that helps firms identify and implement their waste minimization opportunities.

Direct regulatory requirements. Direct requirements are policy directives that have hazardous waste minimization as their primary, and usually exclusive, goal. Five direct requirements for promoting hazardous waste minimization are identified below.

TABLE 11.2   Six Hazardous Waste Minimization Training Opportunities for Local
Government Staff

In-house trainings	Veteran staff can conduct workshops and field observation sessions for less experienced staff. Nearby local governments may have staff experts who would be willing to participate in cross-training programs.
Self-study	Local governments embarking on a technical assistance program need to provide their staffs with time to review the existing waste minimization technical literature. There are a number of published waste minimization assessments from various industries that can be particularly useful.
Other governmental agencies	State and federal agencies are often willing to provide staff experts who can conduct seminars and training sessions.
Outside conferences	Local government staff should take advantage of the numerous national and regional conferences offered on waste minimization.
Consultants	If resources permit, a local government could hire a waste minimization consultant from any one of the many firms specializing in this field. A consultant would conduct technical workshops as well as field observations sessions.
Professional associations	Most local government staff belong to some statewide or national professional association. These associations can be called upon to arrange waste minimization trainings for their members.

Hazardous waste minimization planning requirements.   Hazardous waste minimization planning requirements direct firms to develop a written document that identifies the technically and economically feasible source reduction and recycling measures available, along with an implementation timetable.

Time-tested waste minimization measures.   Where the hazardous waste minimization gains are obvious, basic waste minimization technologies and practices can be identified and mandated by local governments through various local permits and licenses.

Recycling requirements.   Under this local requirement, hazardous waste generators would be required to recycle their hazardous waste when economically and technically feasible.

Employee training requirements.   Local governments can require that firms add a minimization training component to any existing employee education program addressing worker safety, hazardous materials, or emergency response.

Local government internal requirements.   Local governments can issue an internal policy directive requiring that all departments using hazardous chemicals develop and implement a hazardous waste minimization plan.

Indirect regulatory inducements.   Some regulations do not address hazardous waste minimization directly but instead work by expanding or more aggressively enforcing traditional pollution control and hazardous waste regulations. Within this context, waste minimization becomes a cost-effective way to comply with existing regulations that restrict the release of toxic waste and pollutants to air, sewers, waterways, and land. Three indirect regulatory inducements are identified below.

Aggressive enforcement of existing pollution control laws.   Hazardous waste minimization measures are financially attractive, but only for those firms that work within the boundaries of existing environmental laws. That is why aggressive pollution control monitoring and enforcement activities are often necessary for hazardous waste minimization to make economic sense.

Development of more stringent pollution control laws.   Existing pollution control laws, even if rigorously enforced, are currently not stringent enough to prevent the routine release of many chemical wastes into the environment. Local pollution control programs need to examine critically the adequacy of their current emission, discharge, and release limits.

Application of pollution control laws to small and midsize firms.   Smaller firms, often not subject to pollution control laws, can make notable contributions to environmental pollution. Local agencies should evaluate the desirability of placing small and midsize firms under permit.

Positive incentives.   Through creatively using the existing regulatory framework, local governments can offer monetary "carrots" to firms aggressively pursuing waste minimization. Three types of positive incentives are identified below.

Modified fee structures.   Local permit fees that are based on the amount and toxicity of chemicals used, waste produced, or waste shipped off site can reward those firms with active waste minimization programs.

Reduced procedural requirements.   If a firm has significantly reduced its chemical wastes, the frequency of required monitoring, sampling, and other procedures may be able to be safely decreased, saving the company a considerable amount of money.

Reduced fines and penalties.   For firms in violation of pollution control laws, local agencies could reduce fines or other penalties if the firm agrees to pursue waste minimization activities.

11.3   Lead Agencies and Key Players

There are any number of agencies and departments that can serve as the lead for a local waste minimization effort. Table 11.3 lists some of the more common choices. In selecting an appropriate home for the program, local officials should consider the following four points:

1. *A local effort should be structured around a preexisting local government program or function:*   Given the enormous financial pressures on local agencies, this is the only practical way to keep costs down and thereby ensure that the program survives beyond its first year. Thus, an educational program is best handled by those agencies that routinely communicate with the business community. Likewise, a regulatory program that builds upon an existing permit, license, or other programmatic requirement is much more likely to be successful over the long run.

2. *There should be a natural "match" between the scope of the program and the capabilities of the implementing agency:*   As previously mentioned, a technical assistance program needs to be run by an

TABLE 11.3   Potential Lead Agencies for a Local Waste Minimization Effort

Environmental health/health department

Publicly owned treatment works (POTW)

Air pollution control district/agency

Water district/agency

Environmental resources management department

Fire department

Land use planning department

Community development department

Building permits department

Public works department

Local agricultural commission

Public information office

Office of the city manager/chief administrative officer

Any other local agency that regulates hazardous materials/waste or their environmental release agency or department that employs staff with some level of technical expertise. On the other hand, an educational effort may be best coordinated by a department that demonstrates the most commitment to and enthusiasm for the program, regardless of staff expertise. A regulatory program that imposes a waste minimization planning requirement on incoming businesses needs to involve those departments handling the land use, business license, or building permits for new industries.

3. *Regardless of which department or agency is selected as the lead, other relevant departments need to be included in the planning and implementation of the activities from the beginning:* The lead agency provides a coordinating function, but this does not necessarily mean that it implements all of the planned activities. A division of labor with other local government departments, especially those with special expertise, additional resources, or ongoing relationships with businesses, can be more cost-effective than the lead agency trying to do it all by itself.

4. *The lead agency should be able to work closely with local elected officials:* Whenever possible, the agency should invite local elected officials to be part of the program's high-visibility activities (e.g., meetings with local business leaders, press interviews, and public recognition of industries' achievements). Local elected officials bring credibility and status to the program and are key to securing additional resources for program expansion.

11.4   Local Governments in Action

Local governments are just beginning to realize the important role they can play in helping area industries reduce the use of toxic chemicals and the generation of hazardous waste. Compared with other levels of government, they are the closest to the problem and to the solutions. They know their industries and interact with them daily on any number of issues. And, it is the local community that, in the final analysis, has the most to gain or lose from the health of the local economy and environment.

Described below are some local governments that have taken these challenges seriously and instituted local hazardous waste minimization programs. Each program has its own unique features, based on the implementing agency's mission, structure, capabilities, and resources.

Some of the local governments identified below have been able to build comprehensive programs that include educational, technical assistance, and regulatory components. Other programs are more modest, due to resource or other constraints. Many efforts are confined to wastestreams affecting a single environmental medium (i.e., air, water, or land). This is because most local agencies, like their state and federal counterparts, have authority over and responsibility for only one of the three environmental media. The seven case examples given below start with the highly successful Ventura County program, which is described in more detail in another chapter of this book.

11.4.1 Ventura County, California

Ventura County's Environmental Health Department pioneered local government involvement in hazardous waste minimization. Established in 1984, Ventura's program provides information and on-site technical assistance to area industries as part of local hazardous waste generator inspections. Early on, the county identified 75 companies that generated 95 percent of the hazardous waste destined for landfills. Aided by the county's program, these firms implemented aggressive waste minimization and on-site treatment programs. By 1986, the volume of wastes shipped off site had decreased by an estimated 70 percent. The county found that for every dollar of government money invested, industry saved $67. The county now requires that incoming firms that intend to generate hazardous waste must submit waste minimization plans to the Environmental Health Department before they can receive licenses to operate. Ventura County also offers generators a 75 percent reduction on their hazardous waste fees if they serve as a collection site for recyclable waste oil.

11.4.2 Orange County, California

As of the fall of 1988, Orange County Sanitation Districts dedicated one engineer half-time to waste minimization and source control issues. (The districts own and operate two wastewater treatment plants that, in combination, process 250 million gallons of wastewater a day.) The districts offer waste minimization consultations, seminars, and library resources to their industrial wastewater discharge permittees. They may also integrate waste minimization consultations directly into enforcement actions brought against companies violating sewer discharge standards. The waste reduction engineer will also be critically evaluating the air emissions and sludge produced by the POTW (publicly owned treatment works) itself.

11.4.3 Los Angeles, California

The Los Angeles City Fire Department provides educational materials on waste minimization to companies as part of its routine hazardous materials inspections. At a minimum, 8000 businesses will be reached over each two-year inspection cycle. And the Los Angeles City Board of Public Works recently implemented a comprehensive waste minimization effort that focuses on the city's most significant hazardous waste generators. In addition to training all line wastewater treatment inspectors in basic waste minimization technical skills, the board of public works intends to develop a core group of staff waste minimization specialists. The city will continue to aggressively enforce its local wastewater discharge limits, while providing waste minimization technical assistance to its permittees. The city will also be conducting an internal waste minimization assessment to evaluate and implement reduction opportunities for itself.

11.4.4 Santa Clara County, California

In a move to protect area groundwater and decrease the need to find sites for large and unpopular hazardous waste management facilities, the Santa Clara Valley Water District awarded a $175,000 hazardous waste reduction grant to the County of Santa Clara. The 18-month project, which began in mid-1989, will offer technical assistance to Silicon Valley firms, establish a clearinghouse, and publicly recognize local industry successes. A number of innovative strategies will be explored, such as involving area financial institutions in waste minimization loan arrangements, requiring local firms to submit waste minimization progress reports, and integrating waste minimization into local inspection protocols.

11.4.5 Hayward, California

The City of Hayward has begun to develop a waste minimization program for its industrial hazardous waste generators. The program, jointly staffed by Hayward's fire department and wastewater treatment plant, is guided by a five-person advisory commission which is led by a member of the city council. The Hayward program will take a three-step approach to its waste minimization efforts. The first phase will be devoted to educational activities, with technical assistance and regulatory phases scheduled for implementation in the second and third years.

11.4.6 Raleigh, North Carolina

Seven of North Carolina's POTWs provide waste minimization technical assistance to area industries as a routine part of compliance inspections. The Neuse River Wastewater Treatment Plant, a 40-million-gallon-a-day facility operated by the City of Raleigh, has been in the forefront of this issue for several years. Wherever possible, the Raleigh plant recommends alternative compounds and processes that eliminate potential toxic wastewater discharges at their source. In-process recycling is also encouraged.

11.4.7 Suffolk County, New York

The POTW in Suffolk County, New York, requires businesses to identify waste minimization methods when they apply for a wastewater discharge permit. In some cases, POTW engineers have suggested technologies and practice that reduced the discharge to a point where a permit was no longer necessary.

11.5 How a Local Government Can Set Up a Program

Interested local governments can take the first step toward creating a program by drafting and passing a one-page resolution calling for the development of a local hazardous waste minimization effort. To be most effective, the resolution should be formally adopted by the local government's governing body. The resolution should define waste minimization, enumerate the benefits of the program, formally establish the program in concept, identify the lead department, and provide a structure for initial program implementation. Figure 11.1 provides a model resolution for local governments' use.

The directives regarding program implementation are the most important part of the resolution (see the last four clauses of the Figure 11.1 resolution). Those sections ensure that the resolution goes beyond a general policy statement to create an operating program. Specifically, a resolution should instruct local government lead staff to complete the following five tasks:

1. *Identify specific industries, industrial groupings, or pollutant wastestreams for priority attention:*   Establishing clear priorities is an important part of any program's success. But when resources are limited, as they often are for local governments, priority-setting becomes absolutely essential. Given staff expertise, operational costs, projected revenue, and potential outside resources, which companies, types of industries, or specific wastestreams deserve the most attention? Answering this question—on the front end of an effort—will give local government the tools to "get the most bang for its bucks." It will also help staff to track and document future waste minimization progress.

Each community must decide for itself which subset of industry to focus on. Priorities can be guided by one or more of the following cri-

WHEREAS hazardous waste minimization includes reducing the use of hazardous substances, reducing the generation of hazardous waste at the source, and recycling hazardous waste to reduce pollutant releases to all environmental media; and,

WHEREAS hazardous waste minimization saves businesses money by increasing productivity while reducing hazardous waste management costs, short- and long-term liability, and chemical feedstock costs; and,

WHEREAS hazardous waste minimization protects the health and environment of the community, decreases employee exposure to workplace chemicals, and reduces the need for hazardous waste management facilities; and,

WHEREAS the City/County/District of _____ encourages businesses, where feasible, to employ hazardous waste minimization practices, rather than treat and/or dispose of toxic chemical waste into the land, air, and water;

NOW THEREFORE BE IT RESOLVED that the City/County/District of _____ establishes a hazardous waste minimization program to assist area businesses in reducing their hazardous wastes; and,

BE IT FURTHER RESOLVED that the _____Department/Division serve as the lead agency for this effort.

BE IT FURTHER RESOLVED that, in developing this program, the _____(lead department or division) shall:
• Identify specific hazardous waste generators and pollutant wastestreams for priority attention
• Confer with other local agencies that regulate the same industries, particularly those agencies that have authority over the release of toxic pollutants into the waterways, air, and land
• Evaluate the feasibility of each of the following program options: educational outreach, technical assistance, and regulations

BE IT FURTHER RESOLVED that the _____(lead department or division) submit a proposed work program to this Board by (date) that identifies the hazardous waste minimization activities selected for implementation, along with a timetable and required financial support; and,

BE IT FURTHER RESOLVED that the _____(lead department or legal division) recommend to this Board by (date) any changes or additions to local regulations necessary to implement the hazardous waste minimization program as proposed.

Figure 11.1 Model Resolution for Developing a Local Government Hazardous Waste Minimization Program.

teria: (1) large volumes of hazardous chemicals used or hazardous waste generated; (2) large volumes of waste shipped off site, especially to landfills; (3) troublesome pollutants released to the air, sewers, or waterways; (4) common wastestreams generated; (5) high-hazard chemicals used; and (6) chemicals used in close proximity to residential neighborhoods.

2. *Confer with other local public agencies that regulate the same industries:* Table 11.3, displayed earlier in the chapter, demonstrates that a host of agencies can potentially be involved in hazardous material and waste management issues at the local level. Thus, a local government seriously interested in developing an effective waste

minimization program needs to coordinate its efforts with other relevant local public agencies, both within and outside its jurisdiction. Intergovernmental coordination can (1) prevent unnecessary duplication and fragmentation; (2) avoid confusion within an industry that may be the "beneficiary" of two local agencies' waste minimization efforts; and (3) lead to joint waste minimization programs, reducing costs for both agencies.

Coordination is particularly important among the media-specific local agencies that regulate the release of chemical waste to the air, sewers, waterways, and land. This kind of coordination is the only way that local governments can design programs that take a multimedia approach to waste minimization. This point is more thoroughly discussed in the last part of this chapter.

3. *Evaluate all three program elements (educational outreach, technical assistance, regulatory requirements) for possible inclusion into an overall hazardous waste minimization effort:* By implementing the doable ideas from each of the elements, local governments can develop more comprehensive and cost-effective programs.

4. *Submit a proposed work program to the city/county/district governing board, identifying the hazardous waste minimization activities selected for implementation, along with timetables and financial support required:* This important step pulls together all the program planning work into a final recommendation for governing body action. With regard to financial support issues, some local governments have integrated a hazardous waste minimization program into their ongoing operations without additional direct expense. Others have provided line-item support for the program out of their general operating budgets. If neither of these alternatives is possible, a local government might consider an increase in the appropriate permit fees.

5. *Recommend any changes or additions to local regulations necessary to implement the hazardous waste minimization program as proposed in Task 4, above:* Most regulatory programs require modifications to local ordinances or codes. Due to their voluntary nature, educational and technical assistance programs may not need formal recognition in the local government's enabling ordinances. However, if funds are allocated to these efforts, a local government may want to establish the importance of the program and the funding source for the program in an ordinance or formal resolution. Regardless of the level of a local government's involvement in hazardous waste minimization, inclusion of a brief policy statement in a local code draws attention to the issue and provides the context for future program support.

Overview Information

Development of a Local Hazardous Waste Minimization Program: A Model Process, 1988. Prepared by Exceltech for the California Partnership for Safe Hazardous Waste Management, a project of the county Supervisors' Association of California, 1100 K Street, #101, Sacramento, CA 95814-3941 (916) 441-4011.

Educational outreach programs	Technical assistance programs	Regulatory programs
□ *Low-Cost Ways to Promote Hazardous Waste Minimization: A Resource Guide for Local Governments*, 1988.*	□ *Hazardous Waste Reduction Guidelines for Environmental Health Programs*, 1987.	□ *Minimizing Hazardous Wastes: Regulatory Options for Local Governments*, 1988.*
Developed by the Local Government Commission, 909 12th St., #203, Sacramento, CA 95814. (916) 448-1198.	Developed by Ventura County Environmental Health Dept. Available from the Department of Health Services. See below for address.	Developed by the Local Government Commission.
	□ *Reducing Industrial Toxic Wastes and Discharges: The Role of POTWs*, 1988.*	
	Developed by the Local Government Commission.	

* California residents may order these publications through the California Department of Health Services/Alternative Technology Section, P.O. Box 942732, Sacramento, CA 94234-7320. (916) 324-1807. Out of state residents should order the publications through the Local Government Commission.

Figure 11.2 Local government start-up kit for developing a hazardous waste minimization program.

All of these points are discussed in considerably more detail in the local government references displayed in Figure 11.2. Local government officials, particularly staff, should find these handbooks very useful for program planning and implementation.

11.6 Future Trends: Building Local Multimedia Programs

In California, as elsewhere, local waste minimization programs have been purposely structured around a preexisting local government function. In some cases, the activities are built around a local agency whose mission is to regulate the release of pollutants to a specific environmental medium, for example a local wastewater sewage treatment plant, also known as a POTW. Other programs are run by agencies that do not have responsibility for pollutant releases per se; for example, a fire or land use planning department.

Regardless of where they are located, local government waste minimization programs promote source reduction and recycling over treatment and disposal as the preferred strategy for managing chemical wastes. Thus, these programs discourage the cross-media transfer of pollutants that often accompanies pollution control, environmental pollutant dispersion, and the land disposal of hazardous waste. Yet this is a long way from what needs to occur to foster a net and simultaneous reduction of all hazardous pollutants entering the air, water, or land. This larger and more important goal can be realized only through effective *multimedia* waste minimization efforts.

California's Local Government Commission (LGC) is currently working on a model that attempts to bridge some of these gaps. Right now, a handful of local environmental health departments and POTWs in California have implemented waste minimization programs. This demonstrates that the local media-specific agencies dealing respectively with hazardous waste land disposal and industrial sewer discharges can make significant contributions within the context of their agencies' ongoing mission. Although air pollution control agencies have encouraged and even required industries to reduce their toxic emissions at the source, none of the regional or local California air agencies has formally instituted a waste minimization program.

In those localities where both the environmental health department *and* the POTW have already developed their own waste minimization program, the LGC hopes to encourage the local air pollution control agency to do the same. After individual programs are under way, these three local agencies, which separately regulate toxic releases to the land, sewers, and air, will have an opportunity to design jointly a coordinated, multimedia approach to waste minimization for the same industrial base.

Cooperation among the three media-specific agencies can provide the context for significant environmental gains. Working together, these local agencies can (1) strengthen the enforcement of traditional pollution control regulations to prevent the transfer of chemical waste from one medium to another and (2) develop innovative educational, technical assistance, and regulatory waste minimization programs that take this cross-media problem into consideration.

In its future work, the LGC plans to encourage local agencies to develop multimedia waste minimization programs through a four-step process:

Step 1: Information sharing. Cooperation among the media-specific agencies should begin with information sharing, a sorely needed activity given the fragmented nature of the state and federal toxics regulatory system. Often, regulators in one arena know surprisingly little of the nuts and bolts of the other media-specific programs.

Step 2: Interagency regulatory review. Based on the knowledge gained from the first step, each agency should critically analyze its existing and proposed regulations from several vantage points: (1) Which regulations encourage pollutants to be transferred from the agency's medium of concern to other media? (2) Can the agency's regulations be modified to prevent this cross-media transfer? (3) Can the regulations of another agency be modified to deter the transfer? (4) Which regulations discourage waste minimization practices? (5) Do any regulations directly or indirectly encourage waste minimization?

Step 3: Coordinated actions. With the information collected under the first two steps, the various agencies can plan any number of coordinated activities, from interagency agreements to resource sharing, planned division of labor, and other forms of joint actions. Initially the activities may be informal. For example, waste minimization technical assistance efforts could be organized to utilize staff expertise within the various media-specific programs. Or, suspected violations could be routinely cross-reported to the proper agency. Eventually, the cooperation may become more formal. For example, a firm could be required to be in compliance with all media-specific regulations to obtain a permit from any one agency.

Step 4: Integrated actions. At some future point, multimedia coordination may evolve into a more sophisticated system of joint inspections, integrated waste minimization technical assistance, or integrated enforcement and permitting. Through eliminating redundancy and streamlining regulatory requirements, an integrated system has the potential for significantly cutting program costs and the associated industry fees that support those programs. In Massachusetts, the Department of Environmental Quality Engineering (DEQE) is actively exploring an integrated programs approach for implementation at the state level.

Due to the media-specific nature of state and federal regulations, integrated actions may be difficult for local agencies to implement if the corresponding state agencies are not actively supporting the concept. However, as discussed in Step 3 above, a local multimedia waste minimization effort is not necessarily dependent on programmatic or regulatory integration. Successful multimedia programs can be built upon creative intergovernmental *coordination*, which does not require the participating agencies to alter their basic regulatory structure.

As of late 1989, this multimedia model for California local governments is still in its infancy. More environmental health departments and POTWs need to establish their own waste minimization programs. Existing programs need to be strengthened. Then, in selected areas, these two agencies need to work together before air pollution control agencies are encouraged to join the effort. So much has yet to be done. With a little luck and a lot of work at the local agency level, the LGC hopes that four to six communities in California will have fully operating multimedia waste minimization programs by 1992.

Profile of
State Waste Reduction
Programs

Roger N. Schecter

Director of
North Carolina Pollution Prevention Program
and Chairman of
National Roundtable of
State Waste Reduction Programs

12.1 Introduction

Waste reduction programs can be innovative and provide successful examples of how government can work positively toward economic development and environmental quality. An effective alternative to negative incentives such as regulations and costly treatment options has been documented through the development of state level waste reduction and pollution prevention approaches. Strong support from industries, governmental representatives, citizen groups, and the state legislatures has been instrumental in making state programs effective and in ensuring that the efforts will continue to be successful. The states have been leaders in this effort and have provided the basis for national action at the Congressional level and for the development of programs with the U.S. Environmental Protection Agency (EPA).

The basic goal is to develop and provide effective alternatives to waste treatment and disposal through reduction of pollution at the source of generation. This reduction includes more than just hazardous wastes designated by the Resource Conservation and Recovery Act (RCRA) or wastes generated by small quantity generators. Industrial and governmental activities generate wastewater discharges, air emissions, solid waste, and fugitive chemical releases, as well as regulated hazardous waste. Formalized programs and efforts involving industries and governments which target source reduction and recycling can be most effective. Reduction cannot be approached as a panacea for zero wastes, and a reduction program cannot be successful without a firm commitment to change the traditional pollution control mentality in recognition of reduction options.

Government's role in this regard requires an innovative shift in environmental protection to include positive technical assistance and financial incentives in addition to regulations and enforcement. State, federal, and local government goals of protecting the environment can clearly be met through preventing pollution, and industry's goal of producing a profit can clearly be met by the same approach. State agencies have developed active programs centered on information clearinghouses, technical assistance, training, and financial incentives relating to active waste reduction efforts.

In the past decade, a considerable amount of activity has been seen in several states regarding waste reduction, pollution prevention, and waste minimization. In 1984 the North Carolina Pollution Prevention Program organized the National Roundtable of State Waste Reduction Programs, which held its first meeting in the spring of 1985. The National Roundtable was organized to promote the development of state programs and to exchange technical and general information on waste reduction; it is composed of governmental, university, and public interest groups. Representatives from the EPA, Congress, and national environmental organizations also participate in the Roundtable. Since 1985, Roundtable members have helped establish and expand waste reduction programs in 35 states. The Roundtable has helped shape current policy and national efforts for multimedia waste reduction by working closely with the EPA; the Congressional Office of Technology Assessment; key members of Congress and their staff; Inform, Inc.; the Natural Resources Defense Council; and the Environmental Defense Fund.

12.2   Survey of State Waste Reduction Programs

A national survey of state waste reduction programs was undertaken in May of 1988. The survey provides the first uniform characterization of state level waste reduction efforts specific to source reduction and recycling. The survey was intended to provide background information useful in generally characterizing state programs. While averaging or summarizing the responses provides a snapshot of a "typical" program, individual programs vary in scope, resources, target wastes, agency location, and type of activities. The remainder of this chapter summarizes state responses in general terms and presents specific program profiles for each state waste reduction program.

The information reported here is based on responses from 43 programs in 35 states (as noted later, 7 states have related program activities in 2 or more agencies). For the purpose of the survey, programs categorized themselves as *active, developing,* or *planned.* Generally, *active* programs have been operational for at least one year and have a budget and a staff. *Developing* programs are in the organizing stages, and *planned* programs are projected. The survey defined *waste reduction* as "source reduction and recycling."

12.2.1   Cumulative development of programs

Table 12.1 shows the dramatic increase in state waste reduction ac-

TABLE 12.1   Cumulative Development of Waste Reduction Programs by Years*

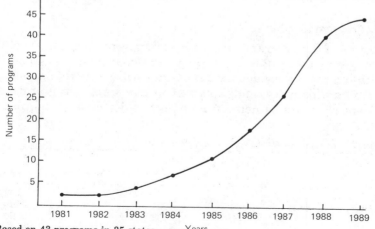

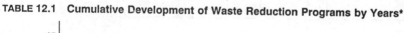

*Based on 43 programs in 35 states.

tivity in this decade. The first programs established in 1981 were primarily industrial materials recycling programs. By 1985 11 programs were established, and by 1987 the total increased to 26. The largest single year's growth was in 1988, when 14 new programs were initiated. At least 3 new programs are planned for 1989. All of these increases together represent a threefold increase in state programs since 1985.

12.2.2 Legislative basis

Between 1987 and 1988, an interesting shift in legislative basis can be noted between active and developing programs. Active programs generally had no specific legislative authority for their waste reduction activities. Active programs with a legislative base were established predominantly through general appropriations bills, general departmental authorities, or broad legislative charges. As shown in Table 12.2, this trend has shifted in developing and planned programs, and 62 percent of these programs report a specific legislative authority. Additionally, of all programs started between 1987 and 1988, almost 70 percent have specific legislation related to waste reduction and program functions.

TABLE 12.2 Legislative Basis for Waste Reduction Program

	Active	Developing and planned	All
Yes	45%	62%	54%
No	55%	38%	46%

12.2.3 Program location

The results of the program survey regarding location are presented as percentages in Table 12.3. Of all the programs responding, 45 percent are within a regulatory agency, 28 percent within another agency, 23 percent in a university, and 5 percent in a nonprofit organization. A significant shift can be noted from active to developing and planned

TABLE 12.3 Location of Waste Reduction Program

	Regulatory agency	Other agency	University	Nonprofit
Active	30%	30%	30%	10%
Developing and planned	60%	25%	15%	0%
All	45%	28%	23%	5%

programs. Active programs are equally distributed over regulatory, other agency, and university categories. A new program is twice as likely to be in a regulatory agency (60 percent) and half as likely to be in a university (15 percent) as compared to active programs. The trend toward RCRA regulatory agency and toward the hazardous/hazardous and solid waste categories described later is probably due to the early EPA program focus on RCRA and the fact that federal funding available during this period was specifically for RCRA regulatory agencies. Federal funding now specifies multimedia program development, and current programs and proposed activities are broadening to address water, air, solid, and hazardous waste.

12.2.4 Program funding and staff

Funding levels for waste reduction programs vary from "not available" to well over $1 million. Table 12.4 shows the averages of the fiscal year budgets and staffing reported in the survey. Active programs show a more stable budget and staff, averaging a $171,000 annual budget and a three-person staff. Developing and planned programs reflect start-up levels averaging a $78,000 annual budget with a staff of two. Active and developing/planned programs show a significant increase in projected funding levels. A 51 percent increase in annual budget to $258,000 is anticipated for active programs, and an even more dramatic fourfold increase is expected for developing and planned programs. Both increases are mostly due to the availability of federal support. Taken as a whole, all programs show an average staff level of 2.5 and an increase in annual budget of 96 percent, from $143,000 to $281,000.

It is interesting to note that several programs did not specify a budget. Overall, 45 percent responded that levels were not available. One

TABLE 12.4 Average Program Resources Specified for Waste Reduction*

	Fiscal year 1988 budget	Proposed fiscal year 1989 budget	Full-time staff
Active[†]	$170,938	$258,200	3.25
Developing and planned[‡]	$ 78,143	$316,000	1.7
All[§]	$142,696	$281,320	2.5

*Waste reduction was defined in the survey as source reduction and recycling.
†25% not available.
‡55% not available.
§45% not available.

out of four active programs did not specify funding, while more than half of the developing/planned programs did not do so.

12.2.5 Funding sources

Table 12.5 shows that the majority of funding for state waste reduction programs is from individual state appropriations. A few programs obtained funding from foundation grants or other sources. Prior to 1988 federal sources were very limited, and only 20 percent of the active states had a mix of state and federal funding for 1988. Developing and planned programs are increasingly being supported with mixes of state and federal funding and are showing dramatic increases in resources, as discussed previously. A few state programs are being initially established with federal funding only. However, taken as a whole, state funding is provided to 95 percent of existing programs.

TABLE 12.5   Percentage Distribution of Funding Sources

	Active	Developing and planned	All
State only	60	65	63
Federal only	0	15	8
State and federal	20	15	18
State, federal, and other	5	5	5
Another mix	15	0	8

12.2.6 Media profile

States were asked to check one or more of four waste categories as targets for waste reduction efforts. While all programs address hazardous waste, the differences in responses appear to be directly related to where each program is located. Table 12.6 shows the percentage of programs checking each media. All programs address hazardous waste, 61 percent indicate they address solid waste, less than half address water, and approximately one-third include air. Table 12.7 shows the actual mix of waste categories for the reduction programs. An equal split (27 percent each) of all programs either checked all categories or checked both the hazardous and solid categories. More than half of all programs responding deal with hazardous and solid waste

TABLE 12.6   Percentage of Programs Addressing Various Waste Categories

	Hazardous	Water	Air	Solid
Active	100	50	45	55
Developing and planned	100	48	29	67
All	100	49	37	61

TABLE 12.7   Media Profile for Reduction Programs

	Active	Developing and planned	All
All media	30%	24%	27%
Hazardous and solid	25%	29%	27%
Hazardous only	25%	24%	24%
Hazardous, water, and air	15%	5%	10%
Hazardous, water, and solid	0%	19%	10%
Hazardous and water	5%	0%	2%

(27 percent) or with hazardous waste only (24 percent). Regarding multimedia, 37 percent of the programs either address all categories (27 percent) or hazardous, water, and air categories (10 percent).

A slight shift is noted in developing and planned programs initiated in the last year. Fewer programs are addressing all categories, and more programs are addressing hazardous only or both hazardous and solid. This focus is probably due to federal funding and the location of new programs in state RCRA regulatory agencies. Over the 1987 to 1988 twelve-month period reported on, less than 25 percent of the programs were "multimedia," and 53 percent were hazardous or hazardous and solid.

12.2.7   Program activities

State programs were given a list of activities and asked to check each activity which was part of the program's efforts. Table 12.8 presents the summary distribution for specified categories. Active programs

TABLE 12.8   Percentage of Programs Indicating Respective Activity

	Active	Developing and planned	All
Compliance assistance	83	65	74
General information	95	95	95
Clearinghouse	63	25	44
General technical assistance	89	75	82
On-site technical assistance	84	55	69
Matching grants	32	10	21
Research/demonstration grants	37	10	23
Loans	21	5	13
Tax incentives	21	20	21
Workshops/training	84	75	79
Public education	68	75	72
Waste exchange	21	20	21

typically checked six to eight separate activities, while developing and planned programs checked four to six activities.

A profile of active programs depicts general information, general technical assistance, and compliance assistance as the top three activities, followed by on-site technical assistance and workshops/training. Active programs are also likely to have financial incentives in the form of matching grants and research grants and, to a lesser degree, loans and tax incentives.

Developing and planned programs mirror active programs in the top two activities, but then the order changes somewhat, to workshops/training, public education, and, to a lesser extent, compliance assistance and on-site technical assistance. These newer programs are less likely to offer financial incentives for waste reduction. Two programs, however, indicated that they were looking at regulatory reform as an incentive for waste reduction.

12.3   Individual Program Profiles

Profiles of individual state waste reduction programs as of May 1988 are presented in Table 12.9, which was compiled from completed survey responses and provided background data for the previous general summary of state programs. Organization of the data in Table 12.9 permits uniform comparison of programs on a limited basis for specific program characteristics.

TABLE 12.9   Profile of State Waste Reduction Programs

	Ala.	Alaska	Ariz.	Ark.	Calif.
Status	Active	Active	Developing	Active	Active
Established	1985	1986	1987	1985	1984
Legislation	No	No	No	No	No
Wastes:					
Hazardous	X	X	X	X	X
Wastewater		X			
Air		X			
Toxics		X		X	
Solid		X		X	
Reduction budget (fiscal year 1988)	$120,000	$65,000	$9,000	NA	$1.6 mil.
Proposed reduction budget (fiscal year 1989)	$125,000	$65,000	$45,000	NA	$1.6 mil.
Staff for reduction	2	2	.25	1	11
Funding:					
State	X		X	X	X
Federal		$28,000	X		
Other	X	$37,000			
Location:					
Regulatory			X		X
Other agency				X	
University	X				
Corporation					
Nonprofit		X			
Other					
Activities:					
Compliance assistance	X	X			X
General information	X	X	X	X	X
Clearinghouse		X			X
General technical assistance	X	X		X	X
On-site technical assistance	X	X		X	X
Matching grants		X	X	X	X
Research/demonstrations	X				X
Loans					
Tax incentives					
Workshops/training	X	X			X
Public education	X	X	X		X
Waste exchange	X		X	X	
Other		X			

NOTE: Survey was conducted by National Roundtable of State Waste Reduction Programs in May 1988.

TABLE 12.9   Profile of State Waste Reduction Programs (*Continued*)

	Conn.	Fl.	Ga.	Idaho	Ill.
Status	Active	Developing	Active	Developing	Active
Established	1987	1988	1983	1988	1985
Legislation	Yes	Yes	No		Yes
Wastes					
Hazardous	X	X	X	X	X
Wastewater	X		X		
Air			X		
Toxics			X		
Solid				X	
Reduction budget (fiscal year 1988)	$95,000	$150,000	$66,750	NA	$350,000
Proposed reduction budget (fiscal year 1989)	$142,150	$250,000	$81,300	NA	$400,00
Staff for reduction	3	4	1	NA	2
Funding					
State	X	$150,000	X	X	X
Federal	X	$100,00			
Other					
Location					
Regulatory		X		X	
Other agency	X				X
University			X		
Corporation					
Nonprofit					
Other					
Activities					
Compliance assistance	X		X	X	X
General information	X	X	X	X	X
Clearinghouse	X	X			X
General technical assistance	X	X	X	X	X
On-site technical assistance			X	X	X
Matching grants	X				
Research/demonstrations			X		
Loans					
Tax incentives	X				
Workshops training		X	X		
Public education	X	X	X		
Waste exchange			X	X	X
Other		X		X	X

[a]Department of Environmental Management, Office of Solid and Hazardous Waste.
[b]Department of Environmental Management, Office of Technical Assistance (see end of table for Purdue University Program).

Ind.[a]	Ind.[b]	Iowa[c]	Kans.	Ky.[d]	Ky.[e]
Developing	Developing	Active	Planned	Developing	Developing
1988	1986	1988		1987	1988
Yes	No	Yes		Yes	Yes
X	X	X	X	X	X
	X		X		
	X				
	X	X			
X	X	X	X	X	X
$78,000	NA	$30,000	NA	$30,000	$75,000
NA	NA	$132,000	NA	NA	$206,000
1	1	1.75	NA	2	1.5
X	X	X		X	$33,000
				X	X
X					
	X				$42,000
		X			
X	X	X	X	X	X
X	X	X			
		X		X	X
	X	X			
	X	X			
	X	X		X	X
		X	X	X	X
		X	X		X
	X	X			X
	X	X			X
		X	X	X	X
X		X	X	X	X

[c]University of Iowa (see end of table for Iowa Department of Natural Resources).
[d]Department of Environmental Protection.
[e]Univertsity of Louisville.

TABLE 12.9   Profile of State Waste Reduction Programs (Continued)

	Md.[f]	Mass.[g]	Mass.[h]	Mich.	Minn.
Status	Planned	Active	Active	Developing	Active
Established	1988	1986	1984	1988	1984
Legislation	Yes	No	Yes	Yes	Yes
Wastes					
Hazardous		X	X	X	X
Wastewater		X	X	X	
Air		X	X		
Toxics		X	X		
Solid		X		X	X
Reduction budget (fiscal year 1988)		$45,000	$200,000	NA	NA
Proposed reduction budget (fiscal year 1989)		$45,000	NA	$400,000	NA
Staff for reduction		1	2	4	4
Funding					
State		X	X	X	X
Federal					X
Other					
Location					
Regulatory		X			
Other agency			X	X	
University					X
Corporation					
Nonprofit					
Other					
Activities					
Compliance assistance			X	X	X
General information		X	X	X	X
Clearinghouse		X	X	X	X
General techncial assistance		X	X	X	X
On-site technical assistance		X	X	X	X
Matching grants		X		X	X
Research demonstration					
Loans					
Tax incentives					
Workshops training		X	X	X	X
Public education		X	X		
Waste exchange					
Other			Reg. Reform		

[f]Legislation passed; organization and program planning in process
[g]Department of Environmental Quality and Engineering
[h]Department of Environmental Management

Mo.	Neb.	N.J.[i]	N.Y.[j]	N.Y.[k]	N.C.[l]
Active	Planned	Planned	Active	Active	Active
1988			1985	1981	1983
No	No	Pending	Yes	Yes	No
X	X	X	X	X	X
		X		X	X
		X		X	X
		X		X	X
X	X			X	X
$33,140	NA	NA	$700,000	$250,000	$400.000
$33,140	NA.	$750,000	$1 mil.	$250,000	$550,000
.75	2	NA	7	5	4
			X	X	X
	X	X			
X					
	X		X		
X					X
				X	
	X	X	X	X	X
X	X	X	X	X	X
	X			X	X
				X	X
	X			X	X
					X
			X		X
				X	
X	X	X			X
	X	X	X	X	X
	X		X	X	X
					X
Reg. Reform		X			

[i]N.J. Department of Environmental Protection and N.J. Hazardous Waste Commission
[j]N.Y. Department of Environmental Conservation
[k]N.Y. Environmental Facilities Corporation
[l]North Carolina Pollution Prevention Program

TABLE 12.9   Profile of State Waste Reduction Programs (*Continued*)

	N.C.[m]	Ohio	Okla.	Ore.[n]	Pa.[o]
Status	Active	Developing	Developing	Developing	Active
Established	1986	1988	1986	1987	1988
Legislation		No	Yes	Yes	No
Wastes					
Hazardous	X	X	X	X	X
Wastewater				X	
Air					
Toxics					
Solid	X	X	X	X	X
Reduction budget (fiscal year 1988)	NA	NA	NA	$125,000	NA
Proposed reduction budget (fiscal year 1989)	NA	$850,000	$150,000	NA	$250,000
Staff for reduction	3	2	NA	1.5	.5
Funding					
State	X	X	X	$75,000	X
Federal	X	X		$50,000	
Other		X			
Location					
Regulatory	X		X	X	X
Other agency		X			
University					
Corporation					
Nonprofit					
Other					
Activities					
Compliance assistance	X	X	X	X	X
General information	X	X	X	X	X
Clearinghouse				X	
General technical assistance	X	X	X	X	X
On-site technical assistance	X	X	X		
Matching grants					
Research/demonstrations					
Loans					
Tax incentives			X	X	
Workshops training	X	X		X	
Public education	X			X	X
Waste exchange	X	X			
Other					

[m]Department of Human Resources, Solid Waste Management Section
[n]Oregon; active solid waste recycling program started in 1973
[o]Pennsylvania Department of Environmental Resources

Pa.[p]	Pa.[q]	R.I.	Tenn.	Tex.	Va.
Active	Active	Deveoping	Active	Developing	Developing
1986	1981	1988	1986	1988	1988
	NA	Yes	Yes	Yes	No
X	X	X	X	X	X
	X	X	X	X	
		X	XX		
	X				
	X			X	
$200,000	NA	NA	$55,000	NA	$80,000
$200,000	NA	NA	$450,000	$189,000	$70,000
2.5	3	1	2	3.5	2
X	X	X	X	X	X
X	X				
X					
			X	X	X
X	X	X	X		
X					
X	X		X	X	
X	X	X	X		X
X	X	X	X		X
X	X	X		X	X
X	X	X	X	X	X
		X	X		
		X	X		
			X		
X	X	X	X	X	X
X		X		X	X
				X	

[p]Pennsylvania Center for Hazardous Material Research
[q]PennTAP, University of Pennsylvania

TABLE 12.9   Profile of State Waste Reduction Programs (*Continued*)

	Wash.	Wis.	Iowa[r]	Ind.[s]
Status	Active	Developing	Developing	Developing
Established	1987	1988	1987	1987
Legislation	Yes	No	Yes	
Wastes				
Hazardous	X	X	X	X
Wastewater	X	X		X
Air	X	X		
Toxics	X	X	X	
Solid	X	X	X	X
Reduction budget (fiscal year 1988)	$125,000	NA	NA	NA
Proposed reduction budget (fiscal year 1989)	$400,000	NA	NA	NA
Staff for reduction	5	.5	.75	2
Funding				
State	X		X	
Federal		X		X
Other				
Location				
Regulatory	X	X		X
Other agency			X	
University				X
Corporation				
Nonprofit				
Other				
Activities				
Compliance assistance	X			X
General information		X	X	X
Clearinghouse	X			X
General technical assistance	X			X
On-site technical assistance	X			X
Matching grants				
Research/demonstrations	X		X	
Loans				
Tax incentives			X	
Workshops trainsing	X	X		X
Public education	X	X	X	X
Waste exchange				
Other				
Reg. reform				

[r]Iowa Department of Natural Resources
[s]Purdue University

The Ventura County Waste Minimization Program

Donald W. Koepp

Ventura County Environmental Health Department
Ventura, California

13.1 Background

California's Hazardous Waste Control Act of 1973 was the first comprehensive hazardous waste control in the United States. It served as a model for other states, as well as for the development of federal hazardous waste control laws. The California State Legislature has designated the Department of Health Services (DHS), Division of Toxic Substance Control, as the responsible agency to carry out state and federal mandates. Up until 1982, local health departments' authority to regulate facilities generating hazardous waste was largely carried out through the use of local nuisance laws. Additionally, some city and county governments used zoning restrictions, or, when issuing land use entitlements, would apply certain conditions and restrictions to the permits.

In 1981, California's hazardous waste control laws were amended to provide local health officers and other local agencies with the authority to enforce state hazardous waste control laws. The law specifically authorized county health officers to enter and inspect any place or vehicle where hazardous waste is transported, generated, treated, stored, or disposed, and to enforce minimum standards and regulations for hazardous waste adopted by the HS.[1] In 1982, the Ventura County Environmental Health Department (EHD) surveyed 100 industries that disposed of hazardous waste and determined that 50 percent were in violation of the state hazardous waste control laws. Violations included illegal disposal practices, manifest violations, and illegal storage practices. Accordingly, the county signed a memorandum of understanding (M.O.U.) with the DHS and proceeded to place under permit all industries which were producing less than 1000 kg of hazardous waste per month and storing it for less than 90 days.

By 1984, the Ventura County EHD was actively inspecting and enforcing state laws in over 1000 businesses. During the period from 1982 to 1983, Ventura County not only proceeded to actively enforce the state hazardous waste control laws and to prosecute violators, but it also developed information to assist industry in complying with the laws and regulations. A number of workshops were provided, educational material was developed, and a compliance manual for generators was developed. The focus of Ventura County's program was to obtain compliance with hazardous waste control laws and, as a related goal, to develop information and education programs to help hazardous waste generators comply with regulations.

As the costs of administering environmental programs and costs of compliance mount, the benefits of reducing chemical use and waste generation at its source become more compelling. Environmental protection efforts emphasizing control and cleanup of hazardous waste *after* it is generated are inadequate. Waste reduction is the most economical and environmentally sound response to the hazardous waste issue today. Furthermore, waste reduction is critical to the prevention of future hazardous waste problems, and it can benefit industry by lowering liability, risks, and costs for regulatory compliance, while minimizing public and environmental exposures.

13.1.1   From controlling generators to controlling waste

In 1983, California passed legislation putting into effect hazardous waste land disposal restrictions and schedules. Additionally, at this time many land disposal sites were being closed in Southern California. More significantly, the land disposal site, where 90 percent of Ventura County's hazardous waste was being disposed of, was threatened with closure. As a result of the land disposal restrictions and the fast-shrinking availability of land disposal options for hazardous waste generators, the Ventura County EHD initiated a Hazardous Waste Minimization Assistance Program to encourage waste reduction and alternative waste management practices. The primary goal of the program was to reduce the quantities of waste being generated and transported from industries in Ventura County for land disposal.

13.1.2   Legislative requirements for waste minimization

The Hazardous and Solid Waste Management Act of 1984 amended the Resource Conservation and Recovery Act (RCRA) and included Congressional mandates for waste minimization. The statutory section applicable to generators of hazardous waste is as follows:

> RCRA Section 3002(b): Effective September 1, 1985 each hazardous waste manifest shall contain a certification that the generator (1) has a program in place to reduce the volume or toxicity of such waste to the degree determined by the generator to be economically practical and (2) that the proposed method of treatment, storage, or disposal is that practicable method currently available to the generator which minimizes the present and future threat to human health and the environment.[2]

The 1984 RCRA amendments also require each generator of hazardous waste to submit a biennial report to the United States Environmental Protection Agency (EPA) which must include a description of the generator's efforts to reduce the volumes and toxicity of wastes generated.

In addition to the federal requirements, California's hazardous waste regulatory program provides a driving force for moving industry toward waste minimization and alternative technologies. The state's land disposal phase-out program, recycling regulations, and waste reduction efforts promote safe handling of hazardous wastes which includes source reduction, recycling, and treatment. Title 22 of the California Administrative Code beginning with Article 12, Section 66763, discourages the disposal of recyclable wastes in California.[3]

13.2 Ventura County Waste Reduction Program

As a result of these new requirements and the unavailability of disposal options for hazardous waste generators, the Ventura County EHD began a program in 1985 to encourage companies to use waste reduction technologies and alternative management practices. Ventura County EHD's approach to waste reduction activities was developed and incorporated into existing hazardous waste generator regulatory program activities. This was done in order to maximize existing resources by building upon expertise and an already existing knowledge base concerning regulations, industry chemical use, and waste generation patterns. Additionally, the existing staff had developed a reputation with local industry of providing accurate and timely information and assisting industry to achieve compliance with state laws. A high level of industrial process knowledge was determined not to be necessary in the initial phases of the program, since the program was developed on the premise that industries know their processes best and, accordingly, are able to select proper waste reduction opportunities when provided with appropriate information. The primary role of the Ventura County EHD personnel was to encourage companies to "get serious" about hazardous waste reduction. The focus was on waste minimization techniques and activities, such as wastestream segregation, inventory controls, spill and leak prevention, and water conservation measures. The program emphasized a "low-tech" approach.

The Ventura County EHD identified six important elements which established the framework of its waste reduction efforts.[4] These are

1. Obtaining waste generation and disposal practices data on Ventura County industries

2. Determining waste reduction potential from the data

3. Obtaining local government support

4. Obtaining industry commitment

5. Training personnel

6. Assisting companies in the development and implementation of waste minimization programs and providing waste reduction information

7. Assisting generators with overcoming barriers to waste reduction

Each of these elements is discussed below.

13.2.1 Data collection and analysis

A major element in the Ventura County program was the ability to obtain data on local waste generators from the DHS manifest information system from 1982 through 1985. This information was used to determine waste generation and disposal patterns for Ventura County industries. In order to project future generation patterns for industry, a questionnaire was also sent to all companies that were currently determined to be relying on land disposal for hazardous waste.

13.2.2 Waste reduction potential

Determining the potential for waste reduction through source reduction, recycling, and treatment is an important step in the development of a local program. This information was determined by comparing wastestream generation data to waste reduction information on practices and technologies.

13.2.3 Local government support

Support from local officials from a policy perspective was extremely important. The premise behind this approach is to formalize the policy within the community and to have a clearly stated policy that focuses on hazardous waste reduction. This association with local government will give support to the program and a clear direction to industry as to the intent of local government.

A general policy statement promoting waste reduction should be made at the county, city, and special district levels, in order to ensure that toxics are not transferred from one media of the environment to another. In Ventura County, the Ventura County Board of Supervisors provided additional support by appropriating $50,000 for the initial phase of the program.

13.2.4 Industry support

Corporate commitment is a prerequisite to accomplishing waste reduction objectives. While plant engineers and managers may identify waste reduction techniques, unless they are given support and resources from "the top," little is likely to happen. To assist in obtaining support for the program, Ventura County obtained corporate commitments by having generators sign agreements that the companies would participate in waste minimization efforts. In Ventura County, letters asking for industry participation were sent to 75 companies, and 69 companies endorsed an agreement to participate.

13.2.5 Hazardous waste reduction personnel training

A successful waste reduction program requires that inspectors are trained to provide waste reduction consultations. The extent and nature of the training will vary, depending upon the backgrounds of the individuals. Additionally, the frequency of further training opportunities will vary depending upon the organization, availability of courses, new technologies, and industry trends.

Understandably, inspectors must be technically trained to be able to deal with very complex compliance management requirements. Perhaps more importantly, they must develop skills in conducting probing interviews with personnel having a wide variety of educational and employment backgrounds. Inspectors must walk a fine line between being supportive to plant supervisors' management of environmental affairs and being assertive in their probing for areas of noncompliance. It is a task requiring both diplomacy and firmness, since the final outcome of the inspection/consultation should be voluntary compliance and waste reduction.

Specifically, organized in-house training can be extremely beneficial. In addition, consultants can bring a large number of different experiences to a training exercise and can provide the expertise for program start-up, which may be otherwise unavailable to an organization.

13.2.6 Providing assistance to industry

The Ventura County Program provided assistance to industry by:

1. Evaluating, in conjunction with plant personnel, waste minimization opportunities.

2. Reviewing waste reduction plans to assure that changes in waste management practices would not violate state laws.

3. Providing information concerning waste minimization opportunities and information related to regulatory compliance issues (such as names and telephone numbers of businesses who produce similar wastes and may be willing to share information on waste minimization for similar processes, and information concerning statewide and local waste exchange recyclers).

4. Conducting workshops and seminars to provide a forum for government and industry to share ideas and mutual concerns regarding hazardous waste reduction; providing technical information by utilizing local industry examples of successful waste minimization strategies; providing a list of qualified consultants who do work in a specific area (many are more than willing to participate in workshops; in one workshop held in Ventura County there were over 400 participants).

5. Distributing written handouts, vendor's list, posters, checklists, and guidelines.

6. Conducting approximately 50 waste audits at the request of companies and making recommendations on recycling, treatment, and handling alternatives that resulted in immediate volume reductions for many businesses.

13.2.7   Helping industry overcome barriers to waste minimization

As a result of working with industry, the Ventura County EHD identified and helped remove several barriers to waste minimization that many industries faced. The need to obtain permits or variances for permits was an area where EHD staff provided considerable assistance. For example, a particular company needed to store hazardous waste for longer than 90 days in order to accumulate sufficient quantities of material to recycle in an economical way. Accordingly, a permit on treatment, storage, or disposal facility is required in California. While a variance for storage of materials may be feasible, the variance process frequently takes longer than the applicant foresees. The Ventura County EHD had the expertise to assist the company in filling out the paperwork for review by state authorities and to make recommendations to the state permitting personnel based on knowledge of the company. The state would accelerate applications submitted in this way and frequently concurred with EHD staff recommendations concerning a desired variance.

EHD personnel were also effective in dealing with internal organizational barriers to waste minimization with companies, e.g., lack of management policies toward waste minimization or lack of authority at local plants to change corporate management policies that affected waste practices. In these cases, the EHD worked with upper managers to help them realize the value of waste minimization.

13.3 Results

In 1984 there were 1116 companies in Ventura County, generating 66,000 tons of hazardous waste, including drilling muds and contaminated soils, most of which was disposed of at class I landfills. Data collected using DHS manifests showed that approximately 75 companies were producing 95 percent of the waste being generated. In 1985 these 75 companies were contacted and asked to participate in a waste reduction program. Sixty-nine companies agreed to participate, and fifty-five companies submitted waste reduction plans. Waste from the metals industry was the single largest standard industrial code represented, although waste from the agricultural, chemical, and defense industries was also represented. Twenty companies employing 1 to 25 people made up the largest category of participants, while at the large end six companies with greater than 500 employees participated.

Figures 13.1 and 13.2 show that the total Ventura County wastestream was manifested to off-site locations for both 1984 and 1987. Manifested hazardous waste in Ventura County has been reduced from 50,717 tons/year in 1984 (excluding drilling muds and contaminated soils) to 9792 tons for 1987, or approximately an 80 percent reduction. The initial phase of the Ventura County program took place in 1985 to 1986, in which a 74 percent reduction occurred. Figure 13.3 tracks wastestream generation from 1984 through 1987. The

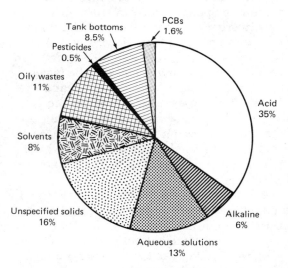

Total tonnage = 50,717

Figure 13.1 Ventura County waste generation comparison: 1984 (drilling muds and contaminated soils not included).

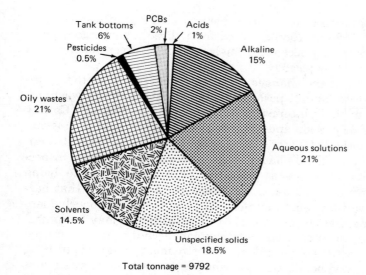

Total tonnage = 9792

Figure 13.2 Ventura County waste generation comparison: 1987 (drilling muds and contaminated soils not included).

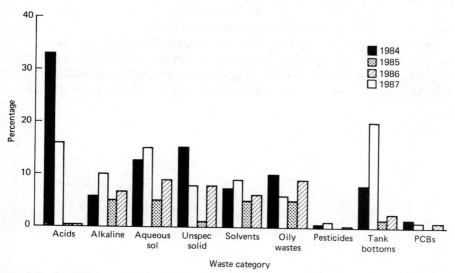

Figure 13.3 Waste production comparison, 1984–87.

largest reduction by wastestream was acids, and this decrease was largely due to the neutralization of acids prior to disposal. Fifty-one companies included in the program used process modification, material substitution, housekeeping segregation, or recycling to reduce waste. It is estimated that about 36 percent of the total wastestream reduction from 1984 to 1986 resulted from these management practices. On-site recycling accounted for less than 1 percent of the total waste reduced for this period

Figure 13.4 shows the changes in waste generation prior to the initiation of the state land disposal restriction or the county's waste reduction program. It is interesting to note that the Ventura County EHD began its inspection and hazardous waste enforcement program in 1982, when only 20,000 tons of hazardous waste was generated annually. During the initial phases of the enforcement program, inspectors encountered significant quantities of hazardous waste being stored for longer than 90 days. EHD inspectors also found that hazardous waste was not being manifested or was being improperly identified on the manifest form. The 1984 disposal figure may be unusually high due to increased enforcement activity and improved identification practices of hazardous wastestreams by industry. In 1982, many industries had not determined whether or not their wastestreams qualified as a hazardous waste. However, by 1984, most industries had completed waste identification. The actual figure for waste manifested for 1986 was approximately 13,000 tons, somewhat less than the 16,500 tons projected. In 1987, there was less than 10,000 tons of hazardous waste manifested from industries in Ventura County.

The reduction reflects the commitment made by industries to institute waste minimization practices that result in source reduction and to utilize existing technologies for on-site treatment. Contributing factors to the 74 percent waste reduction during 1984 to 1986 include

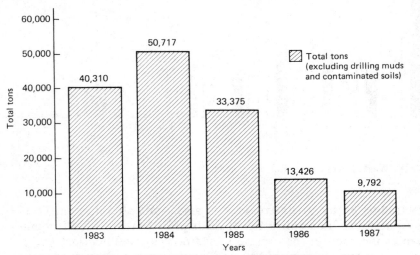

Figure 13.4 Comparison of hazardous waste produced in Ventura County.

high costs for off-site treatment and disposal, governmental regulations, and the EHD's Waste Minimization Program efforts. The EHD's program efforts have been particularly helpful in disseminating information on recycling waste minimization and motivating companies to institute changes. The following are some examples of industry hazardous waste management changes made to minimize hazardous waste generation.

Example 1   This firm blends and applies agricultural chemicals. In 1984 the company shipped 14 tons of pesticide-contaminated rinse waters for disposal. In 1985 the firm installed an experimental wastewater treatment system that consisted of four 55-gallon drums of activated charcoal. The wastewater is pumped through the charcoal, decontaminating the liquid. The contaminated charcoal is disposed of once a year.

PREVIOUS WASTE TOTAL: 14 tons. CURRENT TOTAL: 0.5 tons. REDUCTION: 99 percent.

Example 2   This large consumer product company manufactures toilet paper and diapers. In 1985, the company adopted an operating policy to minimize the generation of hazardous waste. The company mobilized a plant environmental team with representatives in each area to monitor and take steps to avoid the generation of unnecessarily produced waste. The majority of the company's wastes were produced as a result of oil leaks and spills that were corrected through proper maintenance procedures. The company is constantly trying to improve upon hazardous waste management by emphasizing source reduction.

PREVIOUS WASTE TOTAL: 155 tons. CURRENT TOTAL: 50 tons. REDUCTION: 70 percent.

Example 3   This electronics firm manufactures circuit boards. The major reduction of wastes generated has occurred through on-site treatment of plating waste (neutralization and precipitation). As costs for disposal rose, the company increased its on-site treatment capacity and installed filter presses to dry solid wastes. All of the contaminated effluents are now reduced to sludge.

PREVIOUS WASTE TOTAL: 152 tons. CURRENT TOTAL: 92 tons. REDUCTION: 60 percent.

Example 4   This company manufactures silicon discs for use in the electronics industry, and the hydroflouric acid etch process it uses makes on-site treatment difficult. However, to reduce the amount of wastes generated, the company now recycles all solvents, reduces the frequency of acid charges, and has installed a dry etch process in order to use less acid. Improved housekeeping practices have also reduced spillage of chemicals.

PREVIOUS WASTE TOTAL: 15 tons. CURRENT TOTAL: 10 tons. REDUCTION: 33 percent.

13.4   From Waste Generation to Waste Prevention

Since its beginning in 1984, the Ventura County Waste Minimization Program has seen several important developments. The first was the signing into law of AB 2948 (Tanner) in 1986. This law provided for counties and/or cities to prepare hazardous waste management plans. Funding was provided by the State of California. Among the important items required to be included in each plan was an analysis of the jurisdiction's wastestream and the jurisdiction's potential for recycling and reducing hazardous waste. The plan also is required to develop local policies to address hazardous waste issues and is required to be approved by both the cities and the county.[5]

Prior to a plan's submittal to the state for approval, plan guidelines require that the potential for hazardous waste reduction be fully explored before a jurisdiction determines that a storage, treatment, or disposal site is necessary. In addition, plan guidelines require waste reduction in lieu of citing a disposal facility, when possible. The intent of the hazardous waste management plan is to provide the public and decision makers with a document which contains information and policies for management of hazardous waste. Specific waste reduction policies included in the Ventura County plan are as follows:

1. Review manufacturing processes of each new industry locating in Ventura County to ensure maximum waste reduction.

2. Create a clearing house for hazardous waste reduction information specifically for small quantity generators.

3. Establish a common "check-in point" for all businesses which may generate hazardous waste. The purpose would be to conduct preliminary meetings for project developers to inform businesses of waste management requirements.

4. Impose conditions on land use entitlements to reflect the hazardous waste reduction priorities of the plan by applying waste reduction conditions.

These policies, as they are applied to new manufacturing and industrial projects, will strengthen the county's waste reduction program, while providing industry and new businesses in Ventura County with the opportunity to plan for waste minimization activities in conjunction with development projects.

13.4.1 Funding

Funding for the waste reduction program has come from several sources. The first year, a cost of approximately $50,000 was funded by the Ventura County Board of Supervisors. This amount was matched the second year by the DHS. Funding for the program is now provided on a permanent basis through local fee ordinances.[6]

Present California law provides that local jurisdictions may adopt fees to recover the cost of local programs. The State of California had adopted a hazardous waste control fee which it charges all generators who dispose of hazardous waste. However, the state law also provides that local fees charged to hazardous waste generators may be deducted from the permit fee paid to the state. Ventura County has adopted a fee schedule which establishes a fee based upon the amount of waste generated. The fees were established to pay for the cost of administering the waste reduction program,[6] which is approximately $100,000 annually. The purpose of the ordinance is to work with industry to incorporate waste minimization planning into processes as early as possible in the development phases of project proposals. In order to review waste generation production information and obtain waste reduction plans from new industries, Ventura County has adopted a county ordinance requiring waste reduction plans to be reviewed by the county.

The Ventura County EHD's program demonstrates what can be accomplished through cooperation with industry. All of the waste reduction techniques employed have saved companies millions of dollars. Information obtained from 24 companies estimated that their cumulative cost savings would be $2,194,200 annually as a result of their voluntary waste reduction efforts.

13.4.2 Waste minimization program expansion

In California, there are efforts under way to expand waste minimization programs within other local government agencies and sanitation agencies. However, most efforts concerning hazardous waste minimization are being done in Southern California. A statewide survey conducted of county environmental health officials indicated training needs for 173 persons in hazardous waste minimization. Training has started for local health officials in California, and over 200 inspectors have received training in waste minimization as a cooperative effort between government officials and industry.

The Ventura County EHD is continuing its waste reduction program to implement the following objectives:

1. Disseminate information to generators, engineers, and plant operators on waste reduction techniques and alternatives.

2. Work with individual firms to implement waste reduction methodologies.

3. Encourage industries to share information about waste reduction successes.

4. Identify financial resources and incentives for companies needing more capital-intensive technologies.

5. Assist in overcoming regulatory barriers that impede waste reduction efforts.

6. Ensure that new businesses develop waste reduction plans.

7. Determine collection, storage, and transfer station needs of small generators.

The results of the Ventura County program clearly support the value of and need for local agency involvement in hazardous waste reduction efforts.

13.5 References

1. California State Department of Health Services, Hazardous Waste Control Law, excerpts from Health and Safety Code, Division 20, Article 8, Enforcement, Section 25180(a).
2. Deborah A. Hanlon and Donald W. Koepp, "Ventura County Hazardous Waste Minimization Program," *Journal of Environmental Health*, May/June 1987, p. 358.
3. California Code of Regulations, Title 22, Social Security, Division 4, Environmental Health, Office of Administrative Hearings.
4. Ventura County Hazardous Waste Reduction Guidelines for Environmental Health Programs, Ventura County, 1987.
5. Ventura County Draft Hazardous Waste/Materials Management Plan, Policy and Implementation Document, Ventura County, 1988.
6. Ventura County Hazardous Waste Producers' Ordinance, Ventura County, 1988.

Household Hazardous Waste Minimization Program

Linda Giannelli Pratt

Hazardous Materials Management Specialist,
San Diego County Department of Health Services

14.1 Introduction

Public opinion has consistently shown that there is greater concern about hazardous waste than about any other environmental issue. As a consequence of glossy media coverage, *industry* has become synonymous with *hazardous waste*. However, reducing hazardous waste should not be viewed as the responsibility of industry alone. Indeed, achieving a regional goal for hazardous waste reduction must be a shared responsibility of all segments of society: industry, government, and households.

Our society's obsession with household products that clean better, faster, and easier has resulted in an abundance of hazardous chemicals stored in the home. The discarded residual of these household products is household hazardous waste. There is an understandable resistance within community members to acknowledge themselves as hazardous waste generators. Therefore, prior to the development of a successful household hazardous waste minimization program, key issues must be addressed:

1. Clarify what constitutes a household hazardous material and a household hazardous waste.

2. Evaluate the current disposal practices of household hazardous waste.

3. Define which waste groups are most amenable to source reduction and recycling.

A primary goal of any household hazardous materials program should be to effectively present current, relevant information to the community so that workable options can be implemented to reduce the household hazardous waste generated.

14.2   What Is a Household Hazardous Material?

Household hazardous materials are products innocently purchased at the local store which are poisonous, corrosive, flammable, or reactive. Included are routinely used household cleaners, aerosol sprays, automotive supplies, paints and solvents, pool acid, and pesticides. Table 14.1 identifies the hazardous characteristics of typical materials. Ex-

TABLE 14.1   Hazardous Characteristics of Typical Household Materials

Product	Hazardous ingredient	Characteristic
Pesticide	Chlordane	Poisonous
Paint stripper	Methylene chloride	Poisonous
Latex paint	Mercury	Poisonous
Oven cleaner	Sodium hydroxide	Corrosive
Car battery	Acid	Corrosive
Pool acid	Hydrochloric acid	Corrosive
Oil-based paints	Solvents	Flammable
Paint thinners	Hydrocarbons	Flammable
Aerosol cans	Hydrocarbons	Flammable
Copper etching	Picric acid	Reactive
Rodenticide	Zinc phosphate	Reactive

posure to household hazardous materials may cause a variety of acute health effects, such as difficulty breathing, dizziness, nausea, and irritation to eyes or mucous membranes, especially if a product is not used in accordance with label instructions or if two or more products that are used together have incompatible chemical formulations. Ingestion by mouth is a major cause of poisoning in children six years old and under.

14.3 What Is a Household Hazardous Waste?

The U.S. Environmental Protection Agency (EPA) has defined *household hazardous waste* in very much the same way as it defined *hazardous waste*, as stated in the Resource Conservation and Recovery Act (RCRA). Primarily, a household waste is considered to be hazardous if it is

- Ignitable
- Corrosive
- Toxic
- Reactive

Although EPA has defined household hazardous waste, it is not covered under current EPA disposal regulations. This is due, in part, to the overwhelming difficulty of establishing an enforcement program that could maintain community compliance with the regulations.

14.4 What Are Current Disposal Practices?

Recent studies sponsored by EPA report that 0.35 to 0.40 percent of the residential waste that is discarded to landfills qualifies as household hazardous waste. San Diego County residents sent 2.1 million tons of trash to municipal landfills in 1986; of that, according to the EPA percentages, approximately 8400 tons was hazardous waste. The volume being discharged inappropriately into the sewer system and down storm drains has not yet been estimated. One of the critical issues associated with inappropriate disposal practices is confronting the "out of sight, out of mind" attitude of the public. Some examples of the potential consequences of commingling hazardous waste with nonhazardous waste follow.

14.4.1 Routine trash pickup

Typically, trash is picked up at curbside and loaded onto a truck, where the contents are immediately compacted. The compaction can result in a release of material from a broken container, which can result in unsuspecting refuse collectors becoming the victims of dangerous chemical exposure. Pool acid is an example of a typical waste that has caused serious eye injuries, as well as respiratory damage, to refuse collectors.

A second possibility is that the released material does not pose a health threat but instead creates a property damage issue. For example, paint may be unknowingly picked up from one home and hurled into the trash compactor. After compaction, the paint container bursts, and the paint is sprayed onto the driveway of another home. Who is liable to pay for cleaning up the mess?

14.4.2 Municipal landfill

Most often, household hazardous waste is transported without incident to the municipal landfill. Leachates from many of these landfills have been found to contain toxic chemicals in sufficient concentrations to be potentially harmful. Additionally, the equipment operator at the landfill is exposed to the same health risks as previously described for the refuse collector.

14.4.3 Sewer

Continual disposal of household hazardous waste into the sewer or storm drain may result in a gradual deterioration of the collection system. It may even interfere with the biological process at the sewerage treatment plant. Waste that is dumped into an open storm drain channel may become highly accessible to children and animals, thereby posing a threat to public health.

14.5 How Can the Public Utilize Household Hazardous Waste Minimization Techniques?

There is a distinct parallel between what industry can do and what the community can do to reduce the volume of hazardous waste generated. There is a proper hierarchy for waste management practices, and Table 14.2 describes how this can be applied on a household level.

The San Diego County Department of Health Services has administered a Household Hazardous Materials Program since 1983. It is coordinated through the Hazardous Materials Management Division, which serves as a clearinghouse of information for the public. The safe use, storage, and disposal of household hazardous waste is explained by way of telephone inquiries, through an assortment of fact sheets and brochures, and by presentations to interested community associ-

TABLE 14.2   Household Hazardous Waste Management Hierarchy

Source reduction	
Inventory management	Keep track of how quickly each product is being used up and if it is being used correctly.
Prudent purchasing	The cost-per-unit price should not be the primary factor in determining what to buy. If you need only 3 ounces of a product, do not buy a gallon size just because of a lower unit price.
Safe substitutes	When possible, use products made from biodegradable materials.
Recycling	
Material exchange	Contact neighbors, schools, or community organizations that might be able to utilize your surplus materials. Keep in mind that one person's waste is another person's treasure.
Commercial recycling	Typically, items such as waste oil and batteries can be dropped off at designated recycling centers.

ations. However, the focus of the program has shifted, so that it is not merely disposal-oriented. Rather, the message given to the public strongly encourages prudent purchasing, safe substitutes, and recycling.

The program sponsors "collection events" which function as the primary mechanism available to the community to dispose of unwanted household hazardous materials. In 1985, nearly 90 percent of the waste collected was disposed of at designated landfills. Currently, more than 70 percent is recycled or reused. This success is largely due to the three avenues developed to market the usable material. First of all, a material exchange program is set up at the collection events. Household materials such as cans of paint, cleaning materials, automotive lubricants, and soil amendments are displayed and available at no cost for residents. The second avenue for recycling is contacting various organizations that may be able to utilize large volumes of specific material. As an example, soil amendments are provided to the Department of General Services for routine grounds maintenance. In 1987, reuse of these materials saved the department $6700. The third avenue for recycling is utilizing the County Cooperative Auction, which has proved to be very successful in selling the bulk volumes of consolidated water-based paint, which generally comprises 35 percent of the material collected at the collection events. The paint commanded hefty prices at auction, ranging from 50 cents to 2 dollars per gallon!

The Household Hazardous Materials Program has an innovative and effective community education component which promotes safe substitutes and prudent purchasing. Development of this component has largely been the responsibility of the Environmental Health Coalition, which is a subcontractor supervised by county staff. Highlights of the source reduction program follow:

- Curriculums and videotapes have been developed which provide children in grades 4 through 6 with a description of simple waste minimization concepts.
- Three retail distributors have agreed to cooperate in the inventory and labeling campaign for safe substitutes.
- A brochure has been designed which includes a list of safe substitutes for routine household maintenance. An abridged version is shown in Table 14.3.

The community's acceptance and incorporation of these waste minimi-

TABLE 14.3 Safe Substitutes

Product	Alternatives
Aerosol sprays	Choose nonaerosol containers, such as pump spray, roll-on, or squeeze bottles.
All-purpose cleaners	Mix 1 quart of warm water with 1 teaspoon borax, TSP, or liquid soap. Add a squeeze of lemon or a splash of vinegar.
Drain opener	Prevent clogging with a drain strainer on every drain. Pour boiling water down the drain at least once per week. Use a rubber plunger or a metal snake to unclog.
Furniture polish	Dissolve 1 teaspoon lemon oil in 2 pints of mineral oil. Rub toothpaste on wood furniture to remove water stains.
Paint and stains	Latex or water-based paints are the best choice. Cleanup does not require paint thinner.
Paint remover	Use heat gun and scraper to remove paint (be sure to wear proper protective gear). A strong TSP solution (1 pound TSP to 1 gallon hot water) may work well. Brush on, wait 30 minutes, then scrape off.
Chemical fertilizers	Compost is the best soil amendment, and can be made from grass clippings, food scraps, and manure.
Insecticides	Good sanitation in food preparation and eating areas will prevent pests, and they can be sealed out with weather stripping or caulking. For crawling insects use boric acid or silica aerogel in cracks and crevices.
Mothballs	Place cedar chips, dry lavender, or herb sachets in drawers or closets to discourage moths.
Batteries	Old auto batteries can be exchanged when purchasing a new battery or recycled at battery recyclers.
Motor oil	Synthetic motor oil lasts longer than regular motor oil, thereby reducing the amount of oil used. Motor oil can be recycled at participating recycling centers.

zation techniques will gradually occur as awareness of the significant issues increases. It is estimated that less than five percent of San Diego's households are currently participating in the program; consequently, there is a firm commitment to enhance the effectiveness of the community outreach program. With that in mind, the program has set a five-year goal of reducing the volume of household hazardous waste generated by 30 percent. Perhaps this goal is optimistic, but the time has come to push for the changes that must occur so that sound environmental decisions will be made on the home front.

14.6   Further Reading

Dadd, Debra Lynn: *The Nontoxic Home*, Jeremy P. Tarcher, Inc., Los Angeles, 1986.

Dadd, Debra Lynn: *Nontoxic and Natural*, Jeremy P. Tarcher, Inc., Los Angeles, 1984.

Lifton, Bernice: *Bugbusters: Getting Rid of Household Pests Without Dangerous Chemicals*, McGraw-Hill, New York, 1985.

Golden Empire Health Planning Center: "Making the Switch: Alternatives to Using Toxic Chemicals in the Home," 1986.

Wallace, Dan (ed.): *The Natural Formula Book for Home and Garden*, Rodale Press, Emmaus, Pa., 1982.

Yepson, R. B. Jr.: *The Encyclopedia of Natural Insect and Disease Control*, Rodale Press, Emmaus Pa., 1985.

Case Studies of Successful Waste Minimization Programs

The General Dynamics Zero Discharge Program

Denny Beroiz

General Dynamics
Pomona, California

15.1 Program Origins

In the first half of 1984, the president of General Dynamics, Oliver C. Boileau, commissioned an internal task team to examine the existing structure and function of the environmental operations at various divisions of the company. The company had 100,000 people employed in over 15 states. Its products included space launch vehicles, aircraft, tactical missiles, gun systems, armored tanks, submarines, marine vessels, communication systems, and building and energy materials. The team members, who were from many different states, were to look at both the technical and managerial aspects of the approach to the environment. By midyear, a formal written report summarized their observations and conclusions.

One of the major findings that required attention was the diverse way in which the 11 divisions organized the function. In some plants, a facilities engineer, possibly with a background in water treatment or whose part-time duties were energy management, was given the title of "Environmental Engineer," and his or her duties were mainly limited to filing permit applications and generating annual reports required by regulatory agencies. Other locations added these duties into the industrial health and safety function, under human resources. Frequently, this approach did nothing more than supply a single point of contact for the agencies. The safety engineer or industrial hygienist had to coordinate with other departments to get results. Some operating divisions placed the duties on the process engineering, engineering, or quality assurance departments. Still other divisions couldn't readily single out a person who had either the responsibilities or the authority to perform in a reactionary mode. No clear evidence was presented that any division had a proactive approach.

15.2 Program Charter

The findings of the investigators prompted Mr. Boileau to advocate a strong risk reduction approach to future management of the environmental activities of the corporation. He personally chartered a new group that he called Environmental Resources Management (ERM). This group would be the focal point at each division and throughout the company for all environmentally related matters. In November 1984, he directed each of the division general managers to handpick a program leader who, in the president's words, would "take immediate action to reduce or eliminate the use of, and the generation, discharge, and disposal of, hazardous materials, to achieve zero discharge from all facilities." After the candidates were reviewed and each general manager's selection was forwarded to the the corporate office for confirmation of the assignment, Mr. Boileau issued a two-page executive memorandum that outlined the extensive changes to policy that the ERM program would bring to the business.

In December 1984, the new recruits from each division, a few of whom had marginal exposure to environmental law or regulation, were brought together to receive their initial briefing and meet the temporary members of the corporate support team. The briefing supplied the basic elements of the challenge, a measure of the depth of the president's personal commitment, and a timetable for accomplishing major objectives. It was clearly established that compliance to *all* statutory and regulatory requirements was assumed to be the barest minimum standard of performance. The immediate program objectives were (1) to eliminate all electrical equipment that had any detectable polychlorinated biphenyls (PCBs) used as dielectric fluid, (2) to eliminate all single-walled or steel underground tanks, and (3) to eliminate the discharge or disposal of hazardous materials. All of these objectives were to be completed by 1988. The representatives, now designated "Division Program Managers," were to report to the new corporate director of ERM and to some member of the division's executive staff. (Initially, the reporting channels at the divisions were dictated by the historical handling of environmental matters, so the program managers at first reported to facilities, operations (production), or human resources.)

Each program manager was given 30 days to measure and report on the quantity of PCB electrical devices, underground tanks, and hazardous discharges through air, water, and solid waste disposal off site. These figures were to be the benchmark or base line above which no operation was to increase. Within 60 days, the program manager had to give a briefing to the division executive staff on all findings of the initial investigative task team, the base-line data, and planned activities that would put the facility at "zero discharge by 1988" as previously defined by the president. Each program manager was strongly encouraged to present a plan that embodied the "personal concepts" of Mr. Boileau. Some of the earliest expressions of those concepts were to design-out the use of hazardous material in General Dynamics' products; get every phase of the business and every department of the division involved in the execution of the ERM program; muster the resources necessary from already existing supplies; and strengthen the company's financial future through current ERM program accomplishments. It was firmly communicated that the final measure of the program's success would appear in the bottom line and in the company's continued competitive leadership, but also that the contribution would probably be detectable past the 1988 deadline.

15.3 Formal and Informal Powers

The need for documenting and expanding many of these concepts became apparent to the nascent group when its members were faced with the complex and often conflicting objectives of everyday activities "back at the plant." The program managers pooled their inputs and submitted a drafted corporate policy that reiterated the executive memorandum and gave duties and authority to different functions within each division to execute portions of the overall environmental plan. This single activity of policy writing set the foundation for the methodology, imperceptible at first examination, that would be repeated hundreds of times over the next four years in all successful ERM endeavors—a delegated or decentralized approach to problem solving that would allow the group as a whole (through its group dynamics) to accept the responsibility/ownership for whatever outcome it produced. It wasn't until two years into the program that the first recognition of this unspoken, unwritten credo would be formally reported during one of the quarterly ERM corporatewide meetings.

By the first quarter of 1985, the division program managers had already fully implemented the unpublished and still unapproved policy. The authority of the ERM program manager was unbounded but needed consensus or matrix acceptance throughout departments to be realized. Regardless of the "host" department for the ERM program manager, the immediate task was to form a committee of departmental managers that would represent key elements of the division's activity. Most commonly, committees were composed of members from facilities, general counsel, human resources, production, material, quality, engineering, and research.

Two rules were enforced with strong actions—no designated committee representative could subdelegate, and the representative had the full authority of the respective executive staff member in committing resources and in decision making. These rules made ERM committee membership, even though it was an added burden to a manager's existing duties, a privilege and prevented the stigma that "this is just another one of those committees" to which personnel frequently get volunteered. This approach also afforded the program manager the rarest of opportunities—hand selection of the busiest and most trusted personnel from each department. These departmental "thoroughbreds" rapidly became the localized spokespersons, the points of contact and disciples for environmental actions. Plantwide acceptance of ERM concepts and objectives were secured by these opinion leaders who were most frequently chosen for the committee because of their recognition and leadership within their peer group. The frequency of assembly was not uniform, but formal weekly meetings were common for the first 18 months. The lack of technical strength in the environmental field was corrected at all divisions by hiring one or two environmental specialists as a staff for the division program manager.

It was imperative that the division committee draw a sharp line of demarcation between regulatory mandates and discretionary business practices. Also, this required every business practice to be more ambitious than the standards set by the government. The multimedia and multi-issue approach of the program gave the division the visibility over a broad range of regulatory programs that could appear to have conflicting objectives. In areas that combined undefined standards and perceived risk, the situations needed clear and simple definitions to be published. The duty fell upon the division committee to debate and resolve these dilemmas. Once a division consensus was established, the proposed company "standard" would frequently be staffed through the other divisions for viewpoints. Many conservative positions were drawn from proposed state or federal legislation that was meeting opposition and wouldn't be implemented for three or four years.

Quarterly corporate reviews held jointly with all division ERM program managers formed the necessary bond between these environmental pioneers. Monthly and annual reviews of the progress toward the three objectives by management at both the division and corporate executive level kept a sense of commitment and urgency in the program. Within the first year, the ERM program became a section of the strategic plan of each division and was given departmental status during the annual capital budgeting process. Approval by ERM of all planned manufacturing processes and all division capital expenditures above $1000 was required during the second year and from then on. The environmental program was soon being asked to become involved in prenegotiation sessions concerning acquisitions and divestitures involving real property.

15.4 Making the Complex Simple

"Zero discharge" was the term used by the president, and he gave his rationale for the term, once. "It's a single target and easy for you to remember," he explained to the group of anxious program managers, who were having a tough time with the large population of negative thinkers who were demanding a defense for the "technically impossible" task. It became apparent that asking for more definition from the president would demonstrate a learning disability on the part of a program manager. Not grasping single, simple-to-remember targets is career-limiting, even in the most enlightened and progressive setting. Zero PCB equipment, zero undesired tanks, and zero inspection discrepancies are clearly attainable. Later in the program's history, the public-at-large and all company employees would be told that "virtually eliminate the adverse environmental impact of our chemical releases through reduction" was the operative phrase. Even with this expanded definition, when the subject is waste or emissions, the program managers never abandon this rule of thumb: "It isn't zero until you can't measure it."

With simple, tough goals on an accelerated timetable, the program managers were faced with developing simple reporting mechanisms. Factors for variations in production levels were ruled out as game-playing. Process allowances for inherent efficiencies were equally rejected as excuse-making. It was argued that state-to-state variations in waste, air, and water created a nonlevel playing field for the different divisions. Some divisions were operating 30- to 40-year-old facilities; the oldest facility had been in operation for a century. Some facilities were practically new or were in the construction phase, and it was argued that these facilities had an advantage over operations at more mature locations. Many divisions were operating sites that were not owned by the company and were subject to the landowners' acceptance of the changes on the property. All product designs are owned by the customers and cannot be changed without concurrence from many disciplines within the customers' system. These approvals normally take 12 months, if accepted as presented.

No computerization of data was expected for the first two years of the program and no existing computer system could be modified to capture the information. The program desperately needed to feed constant, credible visibility to division and corporate decision makers to ensure their prompt response to a perceived need for resource allocation. There was no place for difficult-to-explain or definition-dependent reporting of progress. The key to success was found once again in the group dynamics of the division program managers.

The recognition that there would never be a level field was critical to the emerging agreements. Also, it was foolish for the program managers to argue over the small differences, since those wouldn't matter until everyone was nearly at zero. Optimistically, it was agreed that zero was still four years away and the level playing field would be discussed then. Meanwhile, a 5-ppm standard of PCBs in electrical equipment was set because of analytical interferences in that range. Whether or not a device was owned by the company didn't matter. If a workplace exposure through the most catastrophic scenario imaginable could result from a PCB or PCB-contaminated device, that device was on the list of equipment to eliminate. Tanks were a simpler case: All underground tanks had to be removed and not replaced below ground. The one exception was gasoline, when local ordinances disallowed aboveground installations. Gasoline tanks required double walls and automatic leak-detection units.

As expected, waste and emissions produced some real challenges. All waste requiring a state or federal manifest is reported and would have to be eliminated through ways other than off-site methods. Off-site recycling, treatment, thermal destruction, alternate fuels, landfilling, and injection wells are all considered undesirable management techniques. Only source reduction, in-plant recycling, waste-to-product conversions, and water treatment can be used to claim waste reduction. Therefore, the phrase "weigh the truck" is used throughout the program to mean both the net weight of the waste under manifest and a reminder to keep the program data management simple. Emissions to air and to water are measured, where practical, or estimated from approximate mass balance calculations. The list of contaminants that are to be tracked is updated annually, with the list growing as more materials are regulated by state or federal programs each year.

All wastestreams and emission streams must have corresponding plans for reduction and elimination, with planned projects ranked and categorized by their contribution to the actual reduction in the quantity in a particular chemical stream.

15.5 Rallying the Champions

The largest single task that the program faced was not identified at the start of the program. While many ERM program participants were concentrating and fretting over the technical obstacles that needed to be scaled, acquiring broad-based support and piercing downward barriers within the departments' organizations became the biggest actual hurdle. When localized ownership of the problem and the corresponding fixes were not firmly established at the start of a change, the lack of workplace acceptance created an insidious failure that was difficult to trace or overcome. More than half of the ERM resources had to be focused on a shift in cultural approaches. This was facilitated by the new ERM division committee structure of managers acting as department representatives.

At one aerospace division, each manager was requested to identify reliable people who would work in smaller focus groups. These focus groups would be given a problem or set of problems and requested to fashion organization, policy, and/or manufacturing changes to resolve the issues. These early career individuals were assembled in logical groups, but frequently one employee would serve multiple groups. The groups' information on wastestreams, emission factors, regulations, and objectives was supplied by an ERM staff employee or the program manager. The group makeup minimally included members from five different departments whose responsibilities overlapped in the target area. It became the practice to have an engineer and one supervisor-level person attend the meetings to assure continuity when other priorities drew personnel resources away. The meetings were short and weekly.

Organizational barriers, both vertical and horizontal, were dissolved as group dynamics worked. Keeping the enthusiasm of members over the life of a project that could possibly take two years was supported by use of executive management briefings. Initially, these were conducted by the division ERM program manager. But with a growing confidence, the focus group members were asked to present the project's scale and progress to an audience of the general manager and selected vice presidents. Participation in the environmental program afforded not only the opportunity to develop more people skills and to get something done that had always been blocked by departmental parochialism, but it also allowed for personal career enhancement from the added recognition and visibility. Environmental programming had escaped the trap of being a "necessary evil" by concentrating on the needs of the worker at the operative level. Three examples of their inspired victories follow:

A total facelift was given to a centralized paint department that had six walk-in booths and was responsible for up to a million component parts per year. The changes included water-based, military-accepted primers and coatings; solvent distillation for reuse; proportional paint mixing at the gun handle; electrostatic and hot turbine spray applications; plastic media paint stripping; robotic application cells; and the installation of emergent technology by using ultraviolet (UV) light and ozone in a volatile organic compound (VOC) destruct system that virtually eliminates process and fugitive solvent emissions.

In the machine shop and metal fabrication operations for degreasing, solvents, like 1,1,1-trichloroethane were replaced. Immersion tanks and industrial grade dishwashers, using specialty cleaners and water, eliminated vapor degreasing and cold-solvent cleaning on bare metal. The complete conversion to water-based cutting fluids eliminated the use of petroleum-based cutting oils and coolants.

A printed circuit board fabrication department was able to replace an aging process line with state-of-the-art equipment that had to pass the "zero discharge" criteria. The process line can produce 1½ million square feet of boards per year and discharges less than 10 gal/min of water (including air scrubber and scheduled tank dump water requirements) with the total heavy metal contaminant level below 300 parts per billion. The chemical feeds and sampling are computer-controlled and are conducted from a remote location so that no worker contact with chemicals or the process solution is ever required. The process line operates without the use of rinse tanks. Heavy metal contamination and processing is handled through extraction techniques that do not generate any wastewater treatment sludges but return metallic forms of the elements to be reused.

15.6 The Unavoidable Bottom Line

In 1984, over 2300 PCB devices existed in service at General Dynamics facilities. More than one-third of these were not owned by General Dynamics but met the criteria of having the potential of exposure impact on operations. In 1988, 95 percent were eliminated or replaced, seven divisions were at zero count, and all units will be gone by 1990. The 1984 base line for single-walled or steel underground tanks was over 300. One-third of the tanks were not owned by General Dynamics. Fifty percent of the tanks have been removed. By 1988, six divisions were at zero, and the program plan for total elimination is intact. Overall, the corporation has experienced a 63 percent reduction in hazardous waste shipped off site by year-end 1988, with no on-site storage or disposal.

At the previously noted aerospace products division, all 19 PCB transformers and all 26 underground tanks were eliminated. The division has approximately 3 million square feet of manufacturing facilities, employs 8000 people, and has sales of $1 billion annually. Most facilities were 35 years old and were not the property of General Dynamics. The hazardous waste generation was reduced by 96 percent by the end of 1988, with the residual being thermally destroyed off site. The use of land disposal for hazardous waste was totally eliminated in 1986. It is projected that the division will soon fit the category of a small quantity hazardous waste generator.

The accomplishments with emissions over the same four years at the aerospace division were 50 percent reduction in the quantity of heavy metals in the water discharge and 70 percent reduction in regulated air emissions (including chlorofluorocarbons [CFCs]). Projects are planned and funded to bring those figures up to 90 percent reduction for both water and air emissions.

The original drafted policy was never modified over the life of the ERM program. It is interesting to note that it was formally released for use 17 months after the first meeting of the new division program recruits. Organizationally, the ERM program offices gradually moved out of facilities and human resources and into operations/production. An annual strategic planning session held with four division program managers and the corporate director was key to focusing and reactivating the program's thrust. With the retirement of O. C. Boileau in 1988, the program was not threatened but received renewed sponsorship at the corporate executive staff level.

Chapter

16

Waste Minimization within the Department of Defense

Joseph A. Kaminski

*Office of the Deputy Assistant Secretary of Defense
(Environment)
Washington, D.C.*

16.1   Department of Defense Organization

The Department of Defense (DOD) is a cabinet-level organization of the executive branch of the federal government. Its components are the Office of the Secretary of Defense, the Organization of the Joint Chiefs of Staff, the military departments, and the defense agencies. The 1988 defense budget was $283 billion. Over 1100 installations worldwide support the defense mission. Major repair of weapons systems is accomplished at 40 maintenance depots. Propellant and ordnance associated with weapons systems is produced at another 20 ammunition plants. Together these industrial installations generate about 80 percent of DOD hazardous waste.

A hazardous waste minimization policy has been issued by the Deputy Assistant Secretary of Defense (Environment). Program guidance stresses source reduction but also urges actions that reduce hazardous waste disposal. Implementation of hazardous waste minimization is delegated to the military departments and defense agencies who assign execution responsibilities to various subordinate commands. The military departments have all adopted goals of 50 percent reduction in hazardous waste disposal by 1992. To achieve this, extensive programs are in place. The following cases cover a variety of waste minimization topics from DOD organizations with different missions.

16.2   Plating and Painting Line Modifications for a Large-Vehicle Repair Operation

Tim Garrett and Tony Pollard
Anniston Army Depot

16.2.1   Introduction

The Anniston Army Depot is a major rework facility of the U.S. Army located in Calhoun County, Alabama. The facility occupies 15,000 acres and is a major employer of skilled and semiskilled workers in northeastern Alabama. Often referred to as the "tank rebuild center of the free world," Anniston Army Depot repairs, overhauls, and converts combat vehicles. The largest share of its maintenance workload is concentrated on the M60 tank and the Army's new M1 Abrams main battle tank. Other missions include storage and renovation of ammunition, storage of all the Army's small arms, repair of mortars and recoil rifles, repair of optical and electronic fire control items, repair of milvan shipping containers, and storage and maintenance of several missile systems, including Shillelagh, TOW, and DRAGON.

16.2.2 Hazardous waste generation

Each year, industrial operations at the Anniston Army Depot generate hazardous waste including paint sludge, obsolete ammunition, ash residue from demilitarization processes, sludge from the industrial waste treatment plant, and spent solutions from chemical cleaning and finishing operations. Table 16.1 shows hazardous waste generated by the major industrial operations. The Anniston Army Depot is constantly pursuing the reduction of hazardous waste volumes and toxicities by implementing new techniques for minimization in the industrial processes. These ongoing techniques include recycling/reusing spent solvents and cutting oils, filtration and subsequent reuse of chemical paint stripping compounds, metal plating/finishing process modifications, spray painting sludge reductions, new paint formulations, and segregation of industrial waste treatment plant sludges with subsequent delisting actions.

Metal plating/finishing. The metal plating/finishing operations at Anniston Army Depot include an automatic barrel plating line and a manual plating line. The automatic barrel plating process (Fig. 16.1) consists of cleaning, pickling, plating, and dichromating. Rinsing is performed after each process step with a two-stage rinse after cadmium plating. Originally all rinses were countercurrent overflow systems controlled by conductivity meters. The automatic line is nearly continuously in operation for one to three 8-hour shifts per day at a parts load of 1200 lb/hour.

TABLE 16.1   Anniston Army Depot Hazardous Waste Generations (in kg)

Waste	1985	1986	1987
Painting	231,965	299,266	444,860
Stripping/nonsolvent metal cleaning	294,229	511,514	453,285
Cleaning/degreasing	150,263	98,391	164,661
IWTP* sludge	653,846	1,400,386	1,246,951
Munitions	2,210,837	1,596,064	1,248,761

*Industrial Waste Treatment Plant

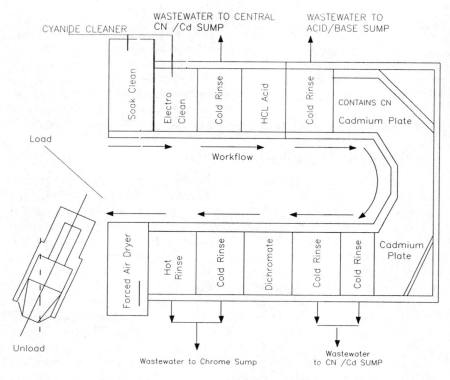

Figure 16.1   Original configuration of automatic barrel cadmium plating machine.

Tests indicated that approximately 0.12 gallon of drag-out was carried by each barrel, resulting in 2.5 lb of cadmium and 10 lb of cyanide per 8-hour shift in the rinse waters. The cadmium- and cyanide-laden rinse waters were fed to a central sump and then to a cyanide destruction unit prior to being discharged to the industrial waste treatment plant. The central sump serves both the automatic and manual plating lines. Total flow, cadmium concentration, and cyanide concentration for the central sump were 10,000 gal/day, 50 ppm, and 100 ppm, respectively.

Modifications to the automatic plating line are shown in Fig. 16.2 and are as follows:

1. The cyanide-based soak clean tank was converted to a reverse current, non-cyanide-based cleaner.

2. The initial rinse station after plating was changed to a stagnant rinse. Once the stagnant rinse becomes saturated with cyanide and cadmium, the water is used for makeup in the plating vats.

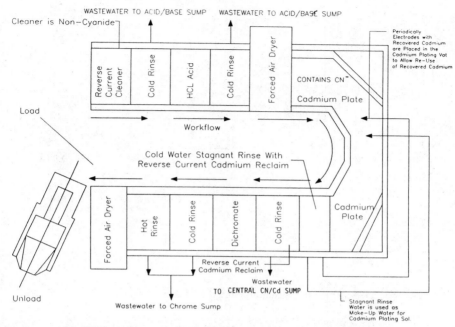

Figure 16.2 Modified configuration of automatic barrel cadmium plating machine.

3. The second rinse station was changed to overflow rinse controlled by a conductivity meter.

4. The stagnant rinse was equipped with a reverse current process enabling the plating of the cadmium in solution onto stainless steel cathodes to be reused in the primary plating process.

Similar changes were made in the manual line as indicated in Figs. 16.3 and 16.4. The manual line operates at a parts load of approximately 800 lb/hour.

These changes in the cadmium plating process have resulted in a recovery of an average of 5 lb of cadmium per day and, more importantly, have reduced the cadmium concentrations by 30 percent, the cyanide concentration by 70 percent, and the flow rate by 40 percent from the central sump. This has reduced waste load, treatment cost, and sludge generation at the industrial waste treatment plant.

Painting wastes. Another area in which the depot is making great strides in the reduction of hazardous waste is in paint sludge generation. In 1986, Anniston switched to a urethane-based camouflage

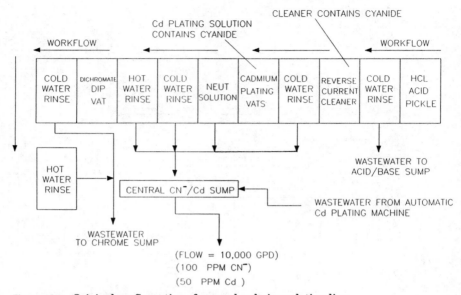

Figure 16.3   Original configuration of manual cadmium plating line.

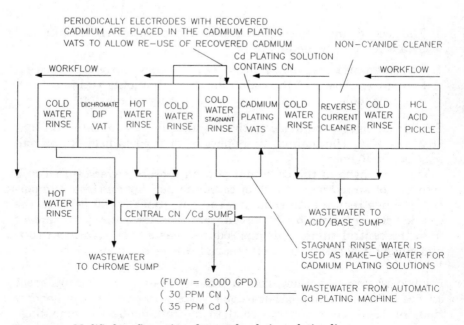

Figure 16.4   Modified configuration of manual cadmium plating line.

paint with a higher solids content. This resulted in a significant increase in paint sludges. Annually, Anniston Army Depot generates approximately 1900 drums of paint sludge at a disposal cost of $330 per drum, resulting in an annual disposal cost of $627,000.

Currently, the Directorate of Maintenance is piloting a system on one of the eight waterfall-type spray paint booths that reduces this volume by 67 percent. The system utilizes a cyclone separator and a paint-detackifying compound to dewater and concentrate the solids fraction of the paint sludge, thus reducing the volume. The detackifier provides a nucleus for overspray particles. A loose suspension is formed, facilitating particle removal by the cyclone. A number of commercial detackifying compounds were tested; however, the most effective compound was a clay-based dust from an on-base paint stripping/blasting operation. Anniston has procured and is currently in the process of installing this system on the seven remaining paint booths in the maintenance operation.

16.2.3 Conclusions

These actions and modifications, along with others, have been effective in lowering wastestream toxicity concentrations and reducing the volume hazardous waste. Anniston Army Depot is committed to even further reductions.

16.3 Hazardous Waste Minimization of Abrasive Blast Media: Addressing Corrosion Control and Environmental Risks

Dale A. Sowell
Department of the Navy
Naval Sea Systems Command

16.3.1 Introduction

The Naval Sea System Command (NAVSEA) is one of five systems commands that provide various ship operating functions for the Office of the Chief Naval Operations. NAVSEA has 13 directorates. The work described in this case was done by personnel of the Engineering Directorate. Within that directorate is the Material and Assurance Engineering Office of Ship Design. Under that office is the Corrosion Control Branch, which is responsible for providing technical direction on corrosion-related issues concerning both new ship design and existing fleet operations. In that branch is the Coatings, Decking and Habitability Section, which is responsible for specifying which abrasive blast media (ABM) is to be used for the removal of paint from ships.

16.3.2 Abrasive blast media (ABM) minimization overview

The abrasive blasting process for cleaning the surfaces of a ship's hull involves impingement of grit (ABM) with high-pressure air to remove paint, rust, and contaminant materials that may adhere to the surface. Proper surface preparation is vital to controlling ship hull corrosion, ultimately extending the life of the vessel. Premature failure of a coating system can result from poor-quality ABM, poor workmanship, inadequate coatings, or improper application. The end results are costly in terms of excessive maintenance and increased overhaul frequency and fuel consumption.

Hazardous waste minimization of ABM is crucial in providing an efficient, productive, and safe environment in which to sandblast. The use of silica sand as an abrasive blast material was previously restricted in recognition that the silica content could be responsible for an increased incidence of the medical condition known as silicosis. Environmental concerns involving ABMs include air-particulate matter, noise, trace metal pollution of navigable waters, and soil and groundwater contamination.

The need for an environmentally and occupationally safe abrasive led to the use of low free-silica slags and mineral grits. Both abrasives offer excellent abrasive cleaning properties and low free-silica content. The use of slag materials, however, does not eliminate all environmental hazards. Mineral slags contain varying concentrations of toxic elements, some of which are carcinogens or suspected carcinogens. The composition of mineral slags may vary depending upon geographic source, the ore and its exact chemical makeup, and production method. These factors will affect the chemical content of the ABM and the severity of the environmental risk posed.

In response to recent health, environmental and productivity concerns, the Navy developed a new ABM specification, MIL-A-22262A. Contaminant limits allowed by MIL-A-22262A are below the standards required to meet a shipyard's water quality discharge permit while assuring abrasive blasting quality. Standards for toxic metals cited in MIL-A-22262A are based on State of California total toxic limit concentrations (TTLC) and soluble toxic limit concentrations (STLC) for identifying a hazardous waste. Table 16.2, excerpted from the specification, lists the limits. Figure 16.5 shows the results of tests on complying and noncomplying ABM, compared to the TTLC standard used to regulate disposal of spent abrasive. Use of the new spec-

TABLE 16.2   MIL-A-22262A(SH): Metals Content Limits for Abrasive Blasting Media

Requirement	Maximum
Soluble Metals Content* (STLC)	
Metals content (soluble), mg/L (type I)	
Antimony and/or its compounds	15
Arsenic and/or its compounds	5.0
Barium and/or its compounds (excluding barite)	100
Beryllium and/or its compounds	0.75
Cadmium and/or its compounds	1.0
Chromium (VI) compounds	5
Chromium and/or chromium (III) compounds	500
Cobalt and/or its compounds	80
Copper and/or its compounds	25
Lead and/or its compounds	5.0
Mercury and/or its compounds	0.2
Molybdenum and/or its compounds	350
Nickel and/or its compounds	20
Selenium and/or its compounds	1.0
Silver and/or its compounds	5
Thallium and/or its compounds	7.0
Vanadium and/or its compounds	24
Zinc and/or its compounds	250
Total Metals Content* (TTLC)	
Metals content (total), weight percent (type I)	
Antimony and/or its compounds	0.05
Arsenic and/or its compounds	0.05
Barium and/or is compounds (excluding barite)	1.0
Beryllium and/or its compounds	0.0075
Cadmium and/or its compounds	0.01
Chromium (VI) compounds	0.05
Chromium and/or chromium (III) compounds	0.25
Cobalt and/or its compounds	0.80
Copper and/or its compounds	0.25
Lead and/or its compounds	0.10
Mercury and/or its compounds	0.002
Molybdenum and/or its compounds	0.35
Nickel and/or its compounds	0.20
Selenium and/or its compounds	0.01
Silver and/or its compounds	0.05
Thallium and/or its compounds	0.07
Vanadium and/or its compounds	0.24
Zinc and/or its compounds	0.50

*Users may acquire to lower than specified limits for toxic components as deemed necessary for local environmental and occupational safety and health requirements.

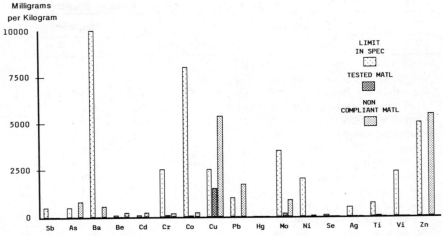

Figure 16.5   Metal content limits.

ification avoids the need to dispose of spent abrasive as hazardous waste merely because disposal limits were exceeded in the virgin material.

16.3.3   History of ABM specifications

An increased awareness and concern for human and environmental risks initially led the Navy to develop ABM specifications. In 1959, the Navy issued MIL-S-22262 for the procurement of quality abrasive blast material that met the needs of the shipyards. MIL-S-22262 addressed the hazardous materials content of ABM as well as the impact on the performance of corrosion control coatings and the effect of these aspects on production and purchasing of ABM.

Suspected health and safety problems posed by ABM led to investigation of possible modifications to MIL-S-22262. In the early 1970s, modifications were accomplished by adding requirements in the Navy Ship's Technical Manual, Chapter 631, to include limitations on low-level radioactive materials and hazardous metals content of the abrasives. The recent issuance of MIL-A-22262A has placed further restrictions on hazardous metals content of the abrasives.

16.3.4   Health and environmental aspects

Occupational safety is the primary concern which requires minimizing toxic chemical constituents of ABM. There are three important variables to consider in minimizing worker exposure to potentially harmful components that may be present in blasting abrasives: (1) the toxic element content of the abrasive, (2) the generation of airborne dust, and (3) the use of personal protective equipment. A basic approach to minimizing hazardous material use and waste production, according to the Chief of Naval Operations, consists of the following:

- Establish limits on the types and quantities of hazardous materials (HM) to be used.
- Control the HM acquisition process.
- Acquire and maintain technical data for each HM.
- Follow proper labeling requirements of HM.
- Train employees handling HM.
- Establish alternatives which minimize hazardous waste (HW) disposal costs.
- Develop means to further reduce HW generation.

In accordance with this guidance, modifications can be made to operational procedures to reduce airborne abrasive dust levels in the workplace. Coarser, harder, larger particles and increased ventilation would lower the dust levels. Rotation of assignments so that a blaster is exposed to abrasive dust for only a part of his or her workday would proportionately reduce the individual's level of exposure. A less hazardous material meeting MIL-A-22262A could be procured. The result would be increased worker safety and a spent ABM that would not be hazardous waste unless contaminated during use.

16.3.5 Acceptance and implementation

The civilian shipbuilding industry and the abrasive blasting industry have relied on the old MIL-S-22262 for ABM and have not established private specifications. They have been reluctant to adopt the modified MIL-A-22262A specification. The industry feels that health, safety, and environmental regulations governing their shipyards can be met by increasing the control methods discussed above without going to procurement of a less hazardous material. The Navy has chosen to use ABM containing lower levels of hazardous materials, making it possible for shipyards to reduce the level of occupational monitoring and work-site controls and facilitate disposal.

Industry reluctance to adopt a more stringent specification is also based on expectation of major cost increases for ABM. The concern is that strict testing requirements could restrict competition by limiting the number of suppliers who can provide abrasives. To determine the availability of ABM qualifying under the new specification, the Navy completed an independent survey. A study examined various unadulterated ABM samples taken from new grit purchased by major private and public shipyards. Included in the testing were chemical and physical properties, metal content, and radioactivity levels. The results of this study indicate that the hazardous materials limits in MIL-A-22262A can be met without reducing the availability or quality of ABM used in the shipbuilding industry.

Navy studies also found that eliminating abrasives that contain unacceptable levels of hazardous constituents also eliminates materials that cause flash rusting of steel and premature failure of the epoxy anticorrosive coating. For these reasons, the Navy has decided to require use of ABM meeting MIL-A-22262A. All eight Naval shipyards are in the process of meeting the requirements for ABM procurement under MIL-A-22262A.

16.3.6 Navy-industry cooperation

NAVSEA is working closely with the shipbuilding industry and the ABM industry and is actively involved with the American Society for Testing and Materials (ASTM) subcommittee F25.02 (Coatings) and the Ship Production Committee SNAME SP-3 (Surface Preparation and Painting) of the National Shipbuilding Research Program. NAVSEA is also actively involved in Steel Structure Painting Council (SSPC) efforts in surface preparation.

In an effort to prepare an industry specification for mineral and slag abrasive media, the SSPC developed SSPC-XABIX. Although the Navy is willing to adopt a commercial specification, SSPC-XABIX does not address the hazardous materials content of ABM. The current version of the specification does not incorporate Navy requirements concerning minimization of hazardous trace elements, as provided for in MIL-A-22262A. SSPC is currently considering a revision to SSPC-XABIX proposed by NAVSEA. At the February 1988 meeting, NAVSEA and SSPC reached an agreement that states: "The Navy has recently agreed to consider adoption of the upcoming SSPC specification as a replacement for MIL-A-22262A(SH) as long as all the requirements of MIL-A-22262A(SH) are reflected in an appendix to the SSPC specification for government work."

16.3.7 Conclusion

The Navy is committed to using an ABM product that allows adequate occupational safety, effectively performs its abrasive blasting function, and minimizes hazardous waste disposal. The Navy, in cooperation with industry, will continue to evaluate and regulate the procurement of ABM and explore new types of ABM and methods of testing.

The adoption of an industry standard that includes all the requirements of MIL-A-22262A(SH) and addresses industrial concerns will enable Naval shipyards to engage in safe and efficient blasting operations, provide reasonable guidelines of ABM content for industry, and minimize hazardous waste.

16.4 Hazardous Chemical Control in a Large Industrial Complex

Robert J. Chabot
San Antonio Air Logistics Center
Kelly AFB, Texas

16.4.1 Introduction

The San Antonio Air Logistics Center (SA-ALC) at Kelly Air Force Base (AFB), San Antonio, Texas, is one of five depot repair facilities in the Air Force Logistics Command (AFLC). The Directorate of Maintenance (MA) within SA-ALC employs over 8000 personnel in the repair and overhaul of large transport aircraft (C-5 Galaxy, C-130 Hercules), strategic bombers (B-52 Stratofortress), and aircraft engines (F100 Jet Engine, TF39 Turbo Fan, T56 Turbo Prop). SA-ALC/MA is greatly dependent upon a wide range of hazardous and nonhazardous chemical materials used in a myriad of processes in a large industrial complex covering 4.4 million square feet of facilities. MA's major processes are electroplating, engine cleaning, paint stripping, aircraft painting, nondestructive inspection, machining, manufacturing, metal plasma spray, foundry operations, and unified fuel control overhaul and testing.

16.4.2 Controlling the use of hazardous chemicals

The Resource Conservation and Recovery Act (RCRA) of 1976, as amended in 1984, imposes extensive "cradle to grave" responsibilities on generators of hazardous waste. To address these responsibilities effectively, a generator must extend management back from generation to use, issuing, stocking, and procuring. Elimination of hazardous material misuse effectively reduces generation of hazardous waste.

An integral part of the hazardous waste minimization program at SA-ALC is the strict control of the processes and chemicals used to accomplish a given set of requirements. The resulting waste from the processes and chemicals chosen must be considered up front in the decision-making process.

The luxury of starting this process from "ground zero" is usually not available. With ongoing processes, the task is first to determine what, where, how, and why chemicals are being used and by whom. It is no small task to ensure that chemicals are purchased, issued, and used in such a way so as to meet process requirements, protect employee health and the environment, and minimize and properly manage resulting waste.

In 1982, concern over waste management practices lead MA to develop a comprehensive waste management program. Over time this program came to include chemical material control and waste minimization. Waste management naturally led MA to chemical material control. The next several sections will describe an important part of MA's entire chemical program, i.e., the control of the purchase and issue of chemical materials.

16.4.3 The objective

Chemical material control program goals must address the requirements of hazard communication standards (employee right-to-know requirements). Chemicals have to be identified as hazardous prior to use and their intended use evaluated and controlled. At SA-ALC, past procedures called for annual workplace surveys but only limited monitoring of chemicals bought throughout the year. We wanted to evaluate the processes to ensure that the right chemicals were being bought and used for the right jobs.

An audit trail was needed to track chemicals to the specific location of use. We wanted to eliminate the unsafe storing and hoarding of chemicals and the borrowing and lending of chemicals among employees who were not authorized or trained. Finally, we wanted to reduce the toxicity and quantity of our waste by better controlling our chemicals.

16.4.4 Implementation

The first step of implementation was re-identifying the shops in MA. A resource control center (RCC)—a cost center—may have as many as 200 personnel who work in several buildings. In most cases, an RCC was broken down to first-level supervisors, thereby identifying chemical usage to that small affected group of 15 to 30 people. The end result was dividing our 96 RCCs into 323 shops, each assigned a unique operation code. Each of the 323 shops was issued a chemical information management notebook. Each book is kept in the immediate work area and contains a site-specific spill plan, the most recent Industrial Hygiene survey for the shop, and most importantly, one AFLC Form 3916 (Fig. 16.6) for each chemical material authorized for use in that shop.

On the top part of AFLC Form 3916, the requester identifies the

CHEMICAL MATERIAL REQUEST/AUTHORIZATION (0094)			1. TYPE REQUEST XX INITIAL ☐ RENEWAL		2. CIMN NO. 308-11
3. ORGN SYMBOL SA-ALC/MATPCE	4. FUNC CODE MTPC9P	5. OPER CODE TPCE3	6. ORGANIZATIONAL TITLE F100/200 EECs & Excitors		
7. BLDG / LOCATION 308, Door 32	8. NSN/LSN 5970 01 021 2623	9. SPECIFICATION Mil-I-46058			10. MO REQ 1

11. NOUN (Include Trade Name, Part No., Mfg. address and unit of issue)	12 JUSTIFICATION (State T.O. callout, mfg callout, work specs, etc.)
HUMISEAL CONFORMAL COATING, P/N IA20 Columbia Chase Corp Woodside, New York 1137	T.O. 6J3-4-102-3, Page 5-19

13. PROCESS (Fully desribe material application ref end items, etc.)

Used to patch the conformal coating on the EEC board assemblies, component connections, and connectors.

14. RESPONSIBLE SUPERVISOR (Name, grade, orgn, phone)	SIGNATURE
Sam Supervisor, WS-12, MATPCE, 200	*Sam Supervisor*

I CERTIFY THE MATERIAL IS REQUIRED AS STATED ABOVE

15. CERTIFYING OFFICIAL (Name, grade, organization, phone)	SIGNATURE
Charles Certify, GS-12, MATEE, 1000	*Charles Certify*

MAINTENANCE ORGANIZATIONS WILL FORWARD TO ENVIRONMENTAL AND INDUSTRIAL SAFETY BRANCH (MAQV), BLDG 308, ROOM 227, EXT 5-4041

16. REMARKS

17. NAME AND TITLE OF REVIEWING OFFICIAL	SIGNATURE
Joe Review, Quality Assurance Specialist	*Joe Review*

FORWARD TO BIOENVIRONMENTAL ENGINEERING SERVICES (EHB), BLDG 305, EXT 5-4544

18. MRC D033 / D002A 7M	19. MSDS Yes, DLA	20. WORK PLACE IDENTIFIER

21. PRIMARY HAZARDOUS MATERIAL(s)

Aromatic solvent, ester solvent, free TDI (less than 1%)

22. PROTECTIVE EQUIPMENT REQUIRED, IF ANY (Check all applicable blocks)

TYPE				SPECIFIC PROTECTION			
SKIN	X	BARRIER CREAM	X	GLOVES	APRON	CLOTHING	RUBBER BOOTS
EYES		SAFETY GLASSES	X	SAFETY GOGGLES	CHEMICAL SPLASH GOGGLES		FACE SHIELD
RESPIRATOR		ORGANIC VAPOR		SPRAY PAINT	METAL FUME / TOXIC DUST		
OTHER (Specify)							

23. EMERGENCY PROCEDURES REQUIRED, IF ANY

Flush eyes or skin with plenty of water, wash with soap and water.

24. REMARKS

Avoid skin and eye contact.
Practice good personal hygiene (handwashing).
Maintain adequate exhaust ventilation.
Recommend finding equivalent with no TDI.

25. NAME AND TITLE OF APPROVING OFFICIAL John Approve Industrial Hygienist	SIGNATURE *John Approve*	26. APPROVAL DATE 8707	27. EXPIRATION DATE 8807

AFLC FORM 3916, OCT 87   PREVIOUS EDITION WILL BE USED   1 - REQUESTING ACTIVITY

Figure 16.6

specific chemical requirement—one form per chemical, per shop. The requester must justify the chemical requirement by attaching a specific call-out for the chemical, such as a technical order/manufacturer's call-out or some other specific justification. The requester must describe the process and the chemical's intended use. The engineering support function for the shop must certify the request.

The form is then forwarded to the Environmental Branch within MA to verify the information, taking into consideration the shop's facilities, personnel training, the process, the material hazard, and the quantity requested. If approved, the request is entered into a computer data-base file and forwarded to Industrial Hygiene. That office enters a material review code, classifying the material's hazard; makes sure that a Material Safety Data Sheet is on hand or ordered; and lists specific health and safety data on the bottom on the form. That office also enters the period of time the form is valid.

Industrial Hygiene keeps a copy for its workplace files and returns the form to MA. The original is placed in a plastic document protector and returned to the shop to be placed in that shop's chemical information management notebook.

16.4.5   Operation

The AFLC Form 3916 is a quick, easily understood reference for employees. It is also a good inspection tool. Local regulation requires an original completed and approved AFLC Form 3916 to be kept in the notebook for every chemical in a shop. We estimate that there are 6000 forms to process and we are 70 percent finished. We expect to finish the first round this year and start on renewals immediately thereafter. Renewals are chemical authorizations which have expired for which the requirement must be reestablished. Renewals will automatically be printed 30 days in advance and sent out to the shops to verify current needs.

If the AFLC Form 3916 is "guts" of SA-ALC/MA's chemical control program, the "teeth" is a Hewlett-Packard HP-3000 computer network which links the MA Environmental Branch with Industrial Hygiene, our supply function, and seven chemical staging areas. This system is called the Chemical Material Management System (CMMS). The data base contains a Master National Stock Number (NSN) file, an Authorization file, and an Actual Purchase File. The Master NSN File provides a unique record specific to each NSN we use. The Authorization file is a computer record of each AFLC Form 3916. The Actual Purchase file is a daily tap transfer of actual purchases made by MA, listed by NSN, shop, and quantity.

The seven computer-linked chemical staging areas are designed to meet all regulatory requirements for storage of hazardous materials and waste. They vary in cost from $60,000 to $191,000, depending on size and features. At present, four are completed and three are scheduled for completion this year.

Figure 16.7 shows how the CMMS and the staging areas link together to ensure that chemicals are given only to authorized and trained personnel. MA requests material through normal supply channels. Supply verifies authorization with the CMMS for an upfront edit on chemical requests. Supply delivers the material to the designated staging area only. The staging area operator verifies the ensuing request for issue of the chemical to the shop against the same authorization list. This provides an additional control check before issue.

Concurrently, the CMMS interfaces with the Maintenance Material Information Control System (MMICS), which contains a computer listing of current personnel with chemical training. Without authorization, the chemical will not be purchased on the front end. On the back end, the material will not be issued without authorization and current training in the chemical's safe use.

All of MA's training was developed in-house and is recorded in the MMICS. There are two general courses and eight specific courses. The specific courses were developed around chemical high-use areas, such as plating, cleaning, and bonding. Course development is ongoing, as needed.

All staging area transactions, whether successful or denied, are recorded daily in the MA Environmental Office. The staging area's computer allows the receipt and issue of chemicals and contains an inventory control program. Comparison of the actual purchase file with

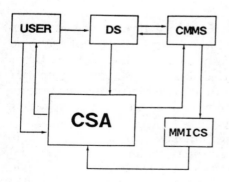

Figure 16.7   Physical and management interface.

staging area transactions gives the Environmental Branch the ability to manage staging area activities.

16.4.6 Learning curve

The staging area designs and the original management concepts were planned in 1983. With construction funding lead times, it took several years for our chemical staging areas to become a reality. SA-ALC/MA has been in the business of forcing its production divisions to justify its chemical requirements before they can purchase chemical materials for 15 months. These 15 months have not been without conflicts and problems, which are worth sharing.

We initially underestimated the time it would take to process the request forms, due mostly to the unforeseen extensive research time required to verify authorizations and enter data. The shops have had a difficult time finding technical data to substantiate their chemical requirements. Often technical data has been absent or incorrect. This whole exercise has greatly improved our technical data. Where warranted, authorization to purchase chemicals was made contingent on submission of appropriate technical data changes.

We experienced initial organizational reluctance in the production and engineering organizations. However, this reluctance was due to the magnitude of the job and not to the idea that it had to be done. Attitudes have improved as the program has progressed.

The initial loading of information into the Master NSN file has been tedious. The microfiche, TOs, Mil-Specs, manufacturer's information, industry standards, etc. necessary to research the material hazards and to verify the shop's requirements are not always easily found. In a large industrial complex such as ours, this is a very labor-intensive undertaking, but its benefits far outweigh its costs.

16.4.7 The future

We are anticipating the future with our chemical management system. Employees already have bar-coded IDs which we will use to access training status. Shops have been issued Chemical Issue Credit Cards—good at your local authorized staging area. I anticipate computer linking of my office with our division engineering and planning functions so that AFLC Form 3916 renewals and new authorizations can be accomplished electronically.

In the more immediate future, with a close control on chemical issues, we expect to limit chemical inventory to that which is essential, similar to a Japanese-style "just in time" system. Where choices exist or become available, the least toxic alternative will be chosen. Upfront process alteration or chemical substitution is our preferred waste minimization technique.

16.4.8 Conclusion

The Environmental Branch within MA not only exercises this computerized control over new chemical materials but also has similar computerized control in its waste management programs. With the Environmental Branch having close control on the purchase and issue of hazardous chemicals and the control of their use and the resulting waste, SA-ALC/MA has a chemical management system with all the loopholes closed. In this day of increasing regulatory requirements and employer liability, a comprehensive chemical control system from purchase to disposal is a necessary part of conducting business.

16.5 In-House Solvent Reclamation Efforts in Air Force Maintenance Operations

Margaret Harris
Warner Robins Air Logistics Center
Robins AFB, Georgia

16.5.1 Introduction

The Warner Robins Air Logistics Center (WR-ALC) at Robins Air Force Base, Georgia, provides worldwide logistics management for the F-15 air superiority fighter and the C-130 and C-140 cargo aircraft, as well as for utility aircraft, helicopters, air-to-air missiles, air-to-ground missiles, a ground-to-air missile, and drone and remotely piloted vehicles. In addition to aircraft management, the center manages the Air Force's fleet of some 140,000 ground vehicles. In addition to depot maintenance responsibility for the above aircraft, WR-ALC is the technical repair center for propellers, life-support equipment, gyroscopes, and airborne electronics. The avionics center at WR-ALC is housed under 11 acres of environmentally controlled, secure facilities. With a workforce of 20,000, WR-ALC is the state's largest industrial complex.

This chapter presents the salient points of the reclamation programs as practiced by Robins AFB and Aerospace Guidance and Metrology Center (AGMC). These programs, while equally successful, are considerably different. These differences stem from the inherent missions of the two organizations. Robins ALC overhauls and repairs aircraft. Large quantities of fairly common solvents are used for cleaning and degreasing in a variety of operations performed during maintenance of the aircraft and its various systems. Examples of these solvents are methyl ethyl ketone, Freon, PD680 or 140 solvent, and 1,1,1-trichlorethane. Robins AFB reclaims a variety of solvents using off-the-shelf equipment, producing products which, while usable, may not meet exact specifications of the original material.

AGMC is the Air Force's principal center for the repair of inertial guidance systems. These systems are precision instruments requiring stringent cleaning with the use of extremely pure solvents. The volume of trichlorotrifluoroethane (Freon) used exceeds that of any other solvent. AGMC's recovery system is designed specifically for the reclamation of Freon to the purity prescribed in the Military Specifications.

16.5.2 Incentive

Most waste minimization efforts are implemented because of a need to comply with current or future environmental regulations, and hence are costly or are perceived as being costly. Reclamation has the appeal of achieving an excellent return on investment and is therefore very attractive. Robins has been operating a successful reclamation program since 1982. Since that time some 50,000 gal of waste solvents have been reclaimed at a material savings in excess of $2 million. In 1987, AGMC reclaimed approximately 100,000 gal of Freon at a material savings of $800,000.

Another indirect benefit of waste minimization is avoidance of disposal cost. For Robins this amounts to a savings of some $200,000 for the period spanning 1982 to 1987. This figure assumes that all waste solvent would have been disposed of in an EPA-approved hazardous waste landfill. No attempt has been made to assign a dollar value to the reduction in long-term liability achieved by preventing waste from crossing facility boundaries, regardless of its destination.

Another benefit of a reclamation program is satisfaction of the 1984 RCRA requirement that all waste generators certify on their manifest, in their annual operating record, and in the biennial report that they have a program in place to reduce the volume, quantity, and toxicity of hazardous waste to the degree determined to be economically practicable.

16.5.3 The AGMC system

AGMC has been reclaiming Freon to the requirements of MIL-C-81302D since 1979, through the use of what is essentially a completely automated continuous distillation plant, with a capacity of 60 gal/hour (see schematic in Fig. 16.8). The waste mixture consists of Freon contaminated with isopropyl alcohol, dioxane, ketones, oil and

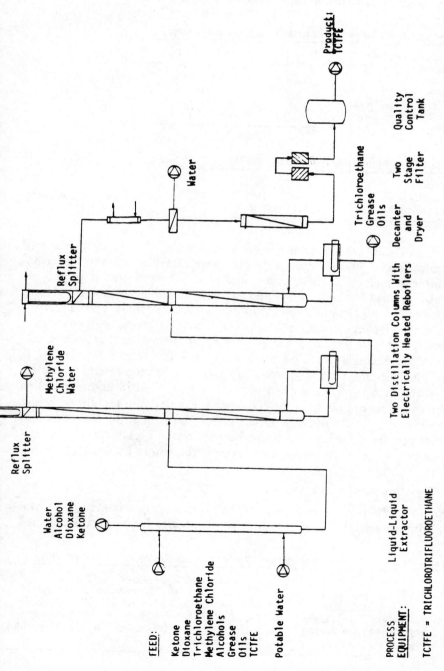

FEED:

Water
Alcohol
Dioxane
Ketone

Ketone
Dioxane
Trichloroethane
Methylene Chloride
Alcohols
Grease
Oils
TCTFE

Potable Water

Reflux
Splitter

Methylene
Chloride
Water

Reflux
Splitter

Water

Trichloroethane
Grease
Oils

Decanter
and
Dryer

Two Stage
Filter

Quality
Control
Tank

Product
TCTFE

PROCESS
EQUIPMENT:

Liquid-Liquid
Extractor

Two Distillation Columns With
Electrically Heated Reboilers

TCTFE = TRICHLOROTRIFLUOROETHANE

Figure 16.8 Artisan TCTFE Recovery System.

293

grease, and chlorinated hydrocarbons such as 1,1,1-trichloroethane and methylene chloride (see Tables 16.3 and 16.4).

TABLE 16.3 Composition of AGMC Contaminated Freon

Alcohols (isopropyl)	100–450 ppm
Ketones (MEK, ACETONE)	100–300 ppm
FC-77	30–360 ppm
Dioxane	20–130 ppm
Chlorinated hydrocarbons	1500–3690 ppm
Nonvolatile materials	0.13–0.40 ppm

TABLE 16.4 Freon Requirement: MIL-C-81302D, Type I

Purity	99.90% (min)
Impurity	0.10% max (1000 ppm)
Nonvolatiles	1.0 ppm max

The AGMC reclamation system is composed of five major pieces of equipment: a liquid-liquid extractor, two distillation columns, a decanter, a molecular sieve dryer, and filters. In the first step, potable water is used in the liquid-liquid extractor to wash the waste solvent, removing water-soluble contaminants such as alcohols. Next is a two-part distillation of washed Freon. The first distillation removes water, methylene chloride, and a Freon/water azeotrope. In the second column the Freon is distilled from the remaining contaminants, 1,1,1-trichloroethane, oil and grease. The distillate is then dried by passing it through a regenerable bed of molecular sieves (in actual practice this step has proved to be unnecessary). Finally, the reclaimed Freon is filtered twice to remove particulates. The first filter is a stainless steel powder filter and the second a membrane-type filter with a 0.2-micron pore size. The reclaimed Freon is then stored in bulk for reuse.

16.5.4 The Robins AFB system

Robins AFB reclaims solvents containing a fairly wide range of contaminants (see Table 16.5) using two centrally located stills manned

TABLE 16.5 Average Composition of Waste Solvents Reclaimed at Robins AFB

92–98%	Freon 113	+	1–5% lube oil & 1–3% water
70–80%	1,1,1-Trichloroethane	+	5–10% 0 + G + 10–30% wax + 5–10% water
85–97%	Isopropyl alcohol	+	1–5% lube oil & 1–5% water & 1–5% nonvolatiles
75–97%	Methyl ethyl ketone	+	1–10% 0 + G & 1–10% water & 1–5% nonvolatiles
70–97%	PD680/Solvent 140	+	1–10% 0 + G & 1–10% water + 1–10% nonvolatiles
50–95%	Paint thinner	+	5–50% paint solids

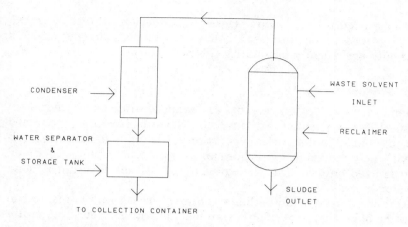

Figure 16.9 Schematic of Solvent Reclaimer.

by full-time operators. Both stills were purchased from Finnish Engineering Co., Erie, Pennsylvania (see schematic in Fig. 16.9). One operates under atmospheric pressure only; the other operates under a vacuum, allowing distillation temperatures as high as 500°F. Each has a capacity of 200 gallons, which can be processed in 5 hours including start-up and shutdown.

Contaminated solvent is fed into the chamber one drum at a time until full. (Similar systems are available for operation in an automatic mode to handle material stored in bulk). The chamber is steam-heated, and the vapors that are produced rise and enter the condenser. The condenser is water-cooled, and it returns the vapors to a liquid state. The clean solvent then flows to a water separator and is then gravity-fed into clean collection drums. The system operates unattended while the batch is being processed. The equipment has proved to be reliable with little downtime other than that required for routine maintenance.

16.5.5 Importance of waste segregation

The success of any reclamation program depends to a large extent on the quality of the segregation of waste solvents. This has been facilitated at Robins by insistence on clear, easily identifiable labeling of all collection containers and by worker indoctrination. With continued emphasis on segregation, the quality of waste solvent has shown a marked improvement over the years. This has served to increase the volume of reclaimable solvent and also to facilitate sale of those drums which are not reclaimed.

Another benefit of the segregation of waste has been the elimination of the need to test drums prior to reclamation. In fact, with the few exceptions to be discussed below, reclaimed material itself is tested only periodically for quality control.

16.5.6 Product quality

As mentioned earlier, few solvents are actually returned to their original specifications using these simple batch spills. However, reclaimed material is used in noncritical operations or is made usable by blending with new material. For example, in gyroscope repair, reclaimed Freon is used only for initial degreasing. All final cleaning is done with new material.

Reclaimed 1,1,1-trichloroethane contains excess water due to the formation of an azeotrope. The reclaimed material is used successfully in vapor degreasers by blending it with new material. However, this practice has led to the stockpiling of a large quantity of reclaimed material. This problem will be solved by the installation of a liquid dryer to remove the excess water, thereby allowing full use of the reclaimed material.

The drier is a single tower dessicant column containing Dow dessicant resin (see schematic in Figure 16.10) capable of drying 1,1,1-trichloroethane with an inlet moisture content of 1000 ppm to a moisture content of less than 75 ppm. The unit will be drained for reactivation when the dessicant is saturated with water. Reactivation is accomplished by passing heated nitrogen over the dessicant bed, stripping off the moisture. The moisture-laden nitrogen then passes out of

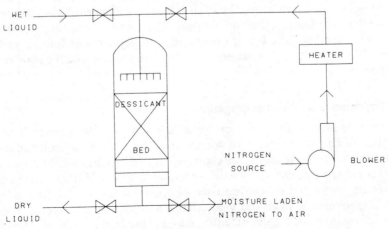

Figure 16.10 Schematic of Drying Apparatus.

the system. At the end of the reactivation time, the heater is turned off and the dessicant bed is cooled by this flow of nitrogen.

Dow dessicant resin (Dowex HCR-W2 sodium form) was chosen over more conventional dessicants such as calcium chloride or silica gel, because it is regenerable at relatively low temperatures and because of its ability to remove water in one pass rather than by recirculation. Also, in general, only water is retained by the dessicant. This is important in order to retain as much of the inhibitors in the reclaimed solvent as possible.

To date, reclaimed 1,1,1-trichloroethane has contained sufficient quantities of acid acceptor inhibitors, probably because of blending with new material prior to reuse. It is possible that depletion of acid acceptors may occur with repeated unblended reuse of dried reclaimed solvent. In order to be prepared for this possibility, 1,2-butylene oxide has been ordered for addition to the reclaimed solvent.

The quality of reclaimed PD680 (140 solvent) has also been unreliable. The initial distillation product entering the condenser is a frothy mixture of the solvent, water, and dirt. The condensed mixture is collected in the separator, which removes the water but not the dirt. Eventually the distillation proceeds smoothly with the production of clean vapor into the condenser. The initial carry-over of dirt, however, contaminates the entire batch. This is thought to be due to bumping caused by the water in the influent. Attempts made to correct this by raising the temperature slowly have failed. An electrically heated system which would allow slower and more uniform distribution of heat may be successful where this steam-heated one is not.

The initial solution to this problem was to filter the product. However, a planned modification of the still should eliminate the need for this additional processing step. This modification involves locating a valve at the entrance to the separator to allow the diversion and separate collection of the initial contaminated solvent.

Robins has also successfully reclaimed polyurethane paint thinners. The new material is comprised of cellosolve acetate 40 percent, toluene 12 percent, MEK 30 percent, and n-butyl acetate 10 percent. The reclaimed material is composed of only toluene and MEK. However the product, which is used only for wipe-down and cleanup of painting equipment, works well and has been readily accepted by the users.

16.5.7 Reclamation waste

Containers containing hazardous waste are not exempt from RCRA requirements, even if reclamation occurs on site. Containers which are used to collect hazardous waste must be labeled with accumulation start dates and contents and be at an unpermitted site for less than 90 days. Still-bottoms derived from the distillation of hazardous waste are a hazardous waste and must be handled appropriately.

16.5.8 Conclusion

Until recently, the success of a waste minimization program was measured almost solely by the volumes of reclaimed solvent, with little consideration to the efficiency with which the solvent was used in the process. With increasing attention to in-process changes to reduce hazardous waste generation, the picture will change dramatically. The successful implementation of a waste minimization plan will invariably mean smaller volumes of waste available for reclamation. Regardless of the changes, however, reclamation will continue to serve as the cornerstone of an efficient waste minimization system.

16.6 Hazardous Property Sales Efforts

Joseph Hoenscheid
Defense Logistics Agency Cameron Station
Alexandria, Virginia

16.6.1 Introduction

Since 1980, the Defense Logistics Agency (DLA) has been assigned the Department of Defense (DOD) mission to dispose of certain commodities of hazardous property. DLA accomplishes this mission through its field office, the Defense Reutilization and Marketing Service, whose 217 Defense Reutilization and Marketing offices (DRMOs) and satellite branches are situated on military installations around the world to act as agents for property disposal. The DRMOs attempt to find property reutilization throughout DOD, other federal agencies, or eligible donees. Should these attempts fail, the property may be sold to eligible buyers or else disposed by a properly licensed commercial contractor.

16.6.2 Sales incentive

The DRMOs are aided in their efforts to sell hazardous property through five region offices, which also perform operational oversight. Traditionally, the region offices have experienced difficulty in reutilization or sales efforts for both used hazardous materials and hazardous waste. As a result, rapid disposal of property through commercial contractors was emphasized during the early years of the program. As the program evolved, costs of commercial contract disposal began to escalate dramatically due to the increasingly stringent EPA hazardous waste disposal regulations and the nation's dwindling waste disposal capacity. As a result, the region offices realized that if greater successes were achieved through sales of hazardous property, not only could DLA reduce disposal costs incurred through use of commercial contractors, but it would also be able to return more sales proceeds back to the DOD.

16.6.3 Actual operation

As a result, sales efforts and successes have increased throughout the DLA disposal network. For example, the concerted effort by the Columbus, Ohio, region office to increase hazardous property sales recently gained DOD recognition through the Defense Productivity Excellence Award Program. This region developed a specific planning document to increase its sales of hazardous property in the Northeastern and Midwestern United States. A hazardous sales team composed of two persons from the Region Merchandising Branch and two from the Sales Branch was formed. The overall goal was to increase sales of hazardous property throughout the Columbus region. To achieve this goal, the team developed these specific objectives:

1. Develop new markets.
2. Provide better sales descriptions within sales catalogs.
3. Increase negotiated sales of hazardous property.

Develop new markets. Developing new markets consisted of the sales team's tracking down new customers for such hazardous commodities as used oil, batteries, and toluene (a cleaning solvent). Prior customer contacts, telephone yellow page directories, and the Thomas Register were the research mainstays. Because of the used status of these commodities, they presented a challenge that required extensive market research. This included searching out and persuading potential customers that used property, such as used oil, could serve the customer's needs as well as the virgin material, and at less cost. The sales team also contacted state waste exchanges and even chambers of commerce. Being able to find buyers for the property was doubly rewarding since it saved the cost of contractual disposal, in addition to removing hazardous property from DOD inventories.

Provide better sales descriptions within sales catalogs. All regions prepare surplus sales solicitations, which require a complete and accurate description of the property being offered for sale. Through the "sealed bid" method, the property is awarded to the highest responsive

IFB 27-8167
IT HAS BEEN DETERMINED THAT THIS PROPERTY IS NO LONGER NEEDED BY THE FEDERAL GOVERNMENT.

SEE INSIDE FRONT COVER FOR NAME, ADDRESS, AND TELEPHONE NUMBER OF PERSONS TO CONTACT
FOR FURTHER INFORMATION AND/OR INSPECTION OF PROPERTY LISTED IN THIS IFB.

PROPERTY IS LOCATED AT PHILADELPHIA, PA

31. BATTERIES: Various mfrs., sizes and types
including nickel cadmium and dry cell. Part
numbers include MS1757/8-2, M81757/11-1, 19804-
08P, MS27307-5P, 369H4540-503, and BA414U. NSNs
include 6140-00-033-9659, 6140-01-037-6794, 6135-
00-125-5256, 6140-01-105-0077 and 6140-01-099-0253.
Inside - Bldg. 912 - Packed and Unpacked - On
pallet (which is included in the weight and sale) -
Used -Poor Condition
Total Cost: $5,548
Est. total wt. 300 lbs.   1 LOT

The following Articles apply if purchased for
regeneration by a battery manufacturer:

AL - Transporting Hazardous Material
BD - Hazardous Property
BE - Government's Right of Surveillance

NOTE: Certification of Intended End-Use appearing
on page 30 applies

RCRA Notice applies, see paragraph 31 on page 21.

In addition to the above, the following Article
applies if purchased by other than a battery
manufacturer for regeneration:

Article BG - RCRA Hazardous Waste

Pre-Award Survey applies, see paragraph 32 on
page 21.

Statement of Intent on pages 33 and 34 must be
completed and submitted with all bids on this
item.

CAUTION: Maintenance Free and Non-Maintenance
          Free Batteries, see paragraph 33 on
          page 21.

32. CYLINDERS, COMPRESSED GAS: Various sizes and ICC
numbers and various DOT numbers, with valves and
caps. Consisting of:
49 ea. - Hallon 1301
12 ea. - CO2
10 ea. - Nitrogen
21 ea. - Oxygen
 8 ea. - Diclorodifluoromethane
 1 ea. - Acetylene
Outside - North Ramp - Loose - Used - Poor
Condition
Total Cost: $43,000
Est. total wt. 10,000 lbs.   1 LOT

The following articles apply:
AH - Compressed Gas Cylinders
BH - Dangerous Property

33. CYLINDERS, COMPRESSED GAS: Carbon dioxide, life
raft, may be empty, charged or partially charged.
Various sizes and ICC numbers with valves. No
provision for protective caps.
Outside - North Ramp - Packed in Tri-Walls and
wood crates on pallets (which are included in the
weight and sale) - Used - Poor Condition
Total Cost: $500
Est. total wt. 11,230 lbs.   1 LOT

The following articles apply:
AH - Compressed Gas Cylinders
BH - Dangerous Property

34. FIRE EXTINGUISHER, CO2 AND DRY CHEMICAL, PKP:
Including various types and sizes, some on carts.
Outside - North Ramp - In Tri-walls, wood crates
and loose (which are included in the weight and
sale) - Used - Poor Condition
Total Cost: $40,000
Est. total wt. 10,160 lbs.   1 LOT

The following articles apply:
AH - Compressed Gas Cylinders
BH - Dangerous Property

35. DISINFECTANT, LIQUID: Consisting of:
120 ea. - Phenolic, Coastal Chemical Co. in 1
          gallon metal cans, packed in cartons on
          pallets. NSN 6840-00-782-2691.
 14 ea. - Tamed Iodine, Wescodyne, in 1 gallon
          plastic jugs, packed in cartons on
          pallets. NSN 6840-00-598-7326.
Inside - Bldg. 995 - Packed - Unused - Poor
Condition
Total Cost: $1036
Est. total wt. 1,000 lbs.   1 LOT

The following Articles apply:
AL - Transporting Hazardous Material
BD - Hazardous Property
BE - Government's Right of Surveillance

RCRA Notice applies, see paragraph 31 on page 21.

Pre-Award Survey applies, see paragraph 32,
on page 21 .

Statement of Intent on pages 33 and 34 must be
completed and submitted with bids on this item.

36. FLOOR FINISH: Eastern Supply Inc., 18% metal
interlock acrylic copolymer, in 55 gallon metal
drums.
Outside - Oil Pad - Loose - Unused - Good
Condition
Total Cost: $3,300
Est. total wt. 15,000 lbs.   30 EACH

The following Articles apply:
AL - Transporting Hazardous Material
BD - Hazardous Property
BE - Government's Right of Surveillance

RCRA Notice applies, see paragraph 31 on page 21.

Pre-Award Survey applies, see paragraph 32,
on page 21.

Statement of Intent on pages 33 and 34 must be
completed and submitted with bids on this item.

Figure 16.11

bidder. In its efforts to sell hazardous property, the Columbus region has prepared its solicitations so that someone interested in purchasing gas cylinders, for example, would know from the solicitation the specific gas, property condition, quantity, weight, applicable environmental regulations, property location, whom to contact for further information, and so on (see Fig. 16.11).

Increase negotiated sales of hazardous property. The Columbus region realized its greatest success through use of the negotiated sale, rather than the sealed bid sale. The negotiated sale enables the sales contracting officer to solicit bids from a limited number of prospective purchasers who are interested in a particular hazardous commodity. In the case of the Columbus region, the negotiated sale effort represented the final attempt to sell property which had failed to sell through the more traditional sealed bid sale, before the government would have to pay for commercial contractor disposal. EPA clearances and other regulations related to hazardous materials often discouraged qualified buyers from bidding on certain property, due to the added cost. Cost avoidance recognized in contract awards was attained through negotiated sales rather than bidding.

Many phone calls, much negotiating savvy, and timely sales contracting administration resulted in successful completion of 81 negotiated sales from January to September 1987, with an estimated commercial disposal cost avoidance of $1.3 million. Sale of used oil was a major success, with 13 negotiated sales accounting for 194,600 gal sold at a cost avoidance of $707,147. Other difficult-to-sell items included 82,000 lb of nickel-cadmium batteries bought by salvage firms and multiple cans of calcium hypochlorite bleach sold through the Thomas Register.

16.6.4 Conclusion

The DLA goal to promote hazardous property reuse and recycling is succeeding. In the fiscal year 1985, the DRMOs sold 38 percent of the hazardous property received from the military services. This increased to 44 percent in 1987. Sale of hazardous property means that unnecessary disposal costs are avoided and environmental resources are conserved.

Waste Minimization in the Petroleum Refining Industry

R. V. Willenbrink

Ashland Oil, Inc.
Ashland, Kentucky

17.1 Background

Ashland Oil, Inc., like other refiners, has historically practiced recycle and reuse. Crude oil has been an expensive raw material in years past, and it is even more expensive today; hence, refiners have always had an economic incentive to recover and recycle waste. However, that incentive is even greater now due to environmental standards which must be met and the high cost of additional treatment capacity and waste disposal. It has always been our policy that before someone recommends new or additional treatment capacity that person (or someone else) must go back to the source (into the process) and determine what improvements might be possible which could reduce waste generation. Improvements range from complete process changes to simple things such as waste segregations and good housekeeping, which in many cases can result in much lower end-of-pipe treatment costs.

In addition to economics, the newest issue which drives waste minimization is the long-term liability associated with on-site or off-site disposal. While this primarily relates to hazardous waste disposal, it may ultimately relate to *all* waste disposal associated with a refining complex.

The incentives for waste minimization have been amplified by the U.S. Environmental Protection Agency's (EPA) imposition of a landfill disposal ban for listed refinery waste unless such waste is treated to EPA specification. The treatment technology required before such waste can continue to be landfilled is incineration (ash disposal) or solvent extraction. An extension of two years from imposition of the landfill ban until August 1990 will be available for refinery listed hazardous waste, with the possibility of two additional one-year extensions because of the lack of prescribed treatment technology capacity. Also, we can expect the list of refinery hazardous wastes to expand. EPA is presently reviewing other wastes generated in a refinery to determine if they should be listed as hazardous. Therefore, waste minimization efforts must be directed at *all* wastestreams in the refinery.

The following are some of the current in-plant systems used to reduce process unit waste generation as well as some of the efforts to reduce the amount of wastes generated by various air pollution control and wastewater treatment systems. Many are commonly used by refiners, with minor variations. Also, you will note that most are rather simple and straightforward. Some applications may be unique to our particular operation, but the point to remember is that waste minimization doesn't always demand advanced technology and large capital costs. In most cases, *you are only limited by your imagination.*

To minimize refinery waste generation, it is necessary to maximize in-plant water reuse and treatment. Of particular concern is the spent phenolic and sulfide caustic streams; foul water containing hydrogen sulfide, phenol, and ammonia; and vacuum tower jet condenser condensate. These streams constitute the bulk of soluble organic waste generated within our refineries. Therefore, their control has a significant effect upon the size and complexity of the final treatment system.

17.2 Contaminant Generation

Phenol, ammonia, and hydrogen sulfide generation occur primarily in the fluid catalytic cracking (FCC) operation. The presence of oxygen in the cracking environment of high temperature produces phenolic compounds by the decomposition of multicyclic aromatics. It is estimated that catalytic crackers can produce 5 to 20 lb of phenol per barrel of charge, depending on the technology used. The majority of the phenol ends up in petroleum products—principally gasoline and diesel fuels. By minimizing the intrusion of air into the reactor system, phenol generation can be greatly reduced. Various methods may be considered, such as stripping the regenerated catalyst with an inert gas. The phenol, H_2S, and NH_3 contaminants generated leave the system with condensate from steam used to strip the FCC products streams, steam used as purge on slide valves and expansion joints, and steam used in the feed nozzles.

Small amounts of similar contaminants are generated in other process equipment and also in product purification. Some products and intermittent streams are washed with caustic to remove mercaptans and some phenols. These streams are then water-washed to remove traces of caustic. Switching to a dilute caustic system allows the majority of phenols which were originally removed to remain with the product. Not only is caustic consumption reduced considerably, but the difficulty in treating high phenolic spent caustic is reduced.

The remaining major refinery contaminant is, of course, oil and its constituents. Of primary concern are those hydrocarbon streams which come in direct contact with water or stripping steam, such as crude oil desalters and fractionation/stripping towers. The next several sections discuss some of the various sources of contaminants and streams susceptible to in-plant reuse and treatment.

17.2.1 Foul water system

Washwaters and steam condensate which have been in contact with hydrocarbon are collected from various unit accumulators and pumped to compressor aftercoolers for use as washwater to prevent corrosion from salt buildup (see Fig. 17.1). (In some refineries there are slight modifications to this initial collection system.) The combined wastewater is then pumped to the main column (FCC product fractionation) vapor line ahead of the condensers. Here, again, the water acts as a wash to remove any ammonium salts and chlorides which may collect in the exchanger bank. The water, now high in phenol, H_2S, and ammonia (foul water), is collected and pumped to the crude column vapor line. Now, in addition to acting as washwater, a fraction of the phenol is absorbed by the hydrocarbons. Tests have shown approximately 50 percent phenol removal by this method. After product separation the water is then discharged to a foul water collection tank where excess hydrocarbons separate out and are recovered. The foul water is then pumped to a conventional foul water stripper for removal of H_2S and NH_3. The off-gas is routed to a sulfur recovery unit where the H_2S is converted to elemental sulfur for resale and the NH_3

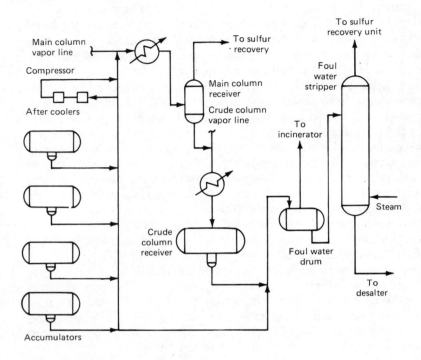

Figure 17.1 Water reuse system.

is thermally destroyed. The foul water stripper effluent is then pumped to the two-stage crude oil desalters for use as washwater, with the additional benefit of phenol removal from the water.

The wastewater from the foul water stripper may be high in phenol. This water, along with other contaminated condensate streams, is then used as desalter wash water. Crude oil prior to downstream processing is mixed with approximately 5 percent water. In a 200,000 barrel/day refinery this is 250 gal/min. The water is used to remove naturally occurring chlorides in the crude oil. The desalter assists in separation of the oil and water. Reusing the water that is high in phenol enables the crude oil to extract up to 95 percent of the phenol, thereby significantly reducing wastewater treatment plant organic load. Additional water required for desalting is supplied by other reuse methods.

Another source of wastewater which can be reused is the vacuum tower jet condensate (see Fig. 17.2). In most cases, we utilize this water after separation of entrained hydrocarbon as desalter makeup water. In other cases, the water is discharged to the foul water stripper

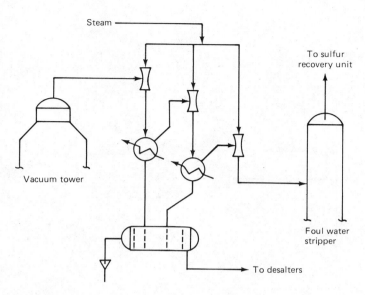

Figure 17.2 Vacuum tower condensate system.

charge drum for subsequent stripping of H₂S and then used as desalter makeup water.

17.2.2 Caustic neutralization

Some refineries are large users of caustic to treat product streams to remove traces of mercaptans and hydrogen sulfide. This caustic must be neutralized before discharge to the wastewater treatment plant (WWTP). A cheap source of acidity is carbon dioxide associated with the fluid catalytic cracker flue gas. By sparging a portion of flue gas through the spent caustic we reduce the pH to acceptable levels for discharge to the WWTP (see Fig. 17.3). The odorous off-gas is incinerated or processed through a gas recovery unit.

17.2.3 Waste oil handling

A major source of oil and oil-water emulsion loss to the refinery sewer system can be the refinery crude oil desalters. If the oil-water emulsion which is purposely formed to allow salt (chloride) removal is not completely resolved in the desalter, an interface (solids stabilized emulsion) forms which builds up and periodically discharges to the oily sewer system through the water draw. The reprocessing of slop oil (API oil separator recovered oil) and the use of stripped foul water can tend to aggravate the emulsion problem. If the slop oil contains too

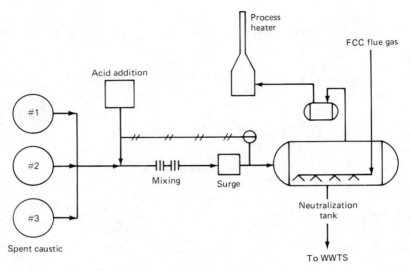

Figure 17.3 Spent caustic disposal.

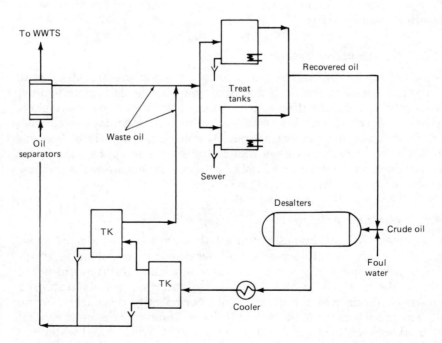

Figure 17.4 Waste oil handling.

much sediment and water or if the ammonia is not sufficiently removed in the foul water stripper, difficult-to-resolve emulsions may be formed. To solve this type of problem requires a complete waste oil handling system within the refinery (see Fig. 17.4). The water from the desalter is cooled to approximately 170°F and sent to two separation tanks in series. The decanted oil is then sent to two heated treating tanks where de-emulsifying chemicals may be added to improve oil-water separation. The treated oil is then returned to the desalter and the water is discharged to the sewer system.

17.3 Process Improvements

Process improvements over time provide obvious opportunities to consider waste minimization. In the mid 1970s we switched from sulfuric acid (H_2SO_4) alkylation to hydrofluoric acid (HF) alkylation. The change was in part due to the high usage of H_2SO_4 associated with the process which required regeneration. The firm available for regenerating the spent acid went out of business leaving us with the options to either ship large volumes of H_2SO_4 several hundred miles for regeneration or change technology. The technology change was chosen because the HF alkylation process uses about 1 percent of the amount of H_2SO_4 previously used, due to the fact that the HF process includes an acid regeneration step.

17.4 Pollution Control System Waste Reduction

Oil recovery and recycling has routinely been practiced for years in refineries by oil separation at the end of pipe (end-of-plant sewer system). Generally it's recognized that the longer the oily wastewater remains in the sewer system, the more difficult the treatment required. Therefore, to minimize this problem, process unit area oil-water separators have been installed, with final oil-water separators at the end of the sewer system.

Oil loss to a sewer system normally comes from hundreds of small sources in a refinery, such as leaking pump seals, valve packing, and sample lines. As individual sources they don't amount to much, but when added up the resultant loss can be very large. As an example, oil recovery/recycling rates at a 200,000 barrel/day refinery were on the order of 2000 barrel/day (84,000 gal/day). In an attempt to reduce that level a person was assigned full-time to survey areas for small leaks and see that appropriate repairs were made in a timely fashion. The results over a year have been amazing. Oil recovery/recycling rates are now in the order of 200 barrel/day (8400 gal/day). As you can see from the example, dedication and attention to seemingly small leaks can have a big overall impact in any waste minimization program. Also, such dedication and attention doesn't require a lot of equipment or cost a lot of money.

Today in Ashland's refineries about 98 percent of the hazardous waste is generated by the WWTP. The WWTP consists of oil separation, equalization, air flotation, and biological oxidation. The hazardous waste in this case is oil separator sludge which is primarily oil-coated solids. Every pound of solid material—primarily dirt that enters the plant sewer system—produces approximately 10 pounds of waste material (oil-solid-water) to handle. The solids mix with the oil in the sewer system, and the resulting mixture will concentrate only to approximately 10 percent without mechanical aid. The obvious way to reduce the amount of oily solids generated is to minimize solids getting into the system. Some very simple ways of doing this are to ensure drainage area erosion control and to use a broom or power sweeper to pick up road dust, catalyst fines, etc. The advice in old sayings "If it doesn't move, paint it" and "If it doesn't move, pave it" works well in this case. A large refiner in the Houston area instituted a routine roadsweeping program recently and found it could pick up several tons of dirt and dust a day. Here again a simple rule: Sweep it up rather than washing it down or letting the rain wash it down. This approach can reduce waste generation significantly.

The other major wastestream from the WWTP was excess biological solids. Although not a hazardous waste, they were a difficult waste to handle since they would only concentrate to 2 to 3 percent solids without mechanical dewatering.

To reduce the volume of waste requiring disposal from the WWTP we installed an automated pressure filtration system to squeeze oil and water from a mixture of oily solids and biological solids. The oil was recovered and reprocessed. To operate a pressure filtration system properly requires the use of a filter precoat consisting of coal fly ash. It is also necessary to precondition the sludge with ferric chloride and lime. All of this adds approximately 40 percent to the dry weight solids ultimately requiring disposal in a hazardous waste landfill. The resultant filter cake is manageable, containing approximately 45 percent moisture, and it passes the EPA leachate test. While such a system adds considerably to the overall dry weight requiring disposal because of preconditioning, the system consistently provides a dry filter cake.

With the addition of the pressure filtration system, the WWTP disposal quantities were still too large. Therefore, to further reduce hazardous waste the biosystem solids were segregated from the oil-separator solids. This required the installation of a second pressure filtration system at the cost of approximately $750,000. Although the second system was an expensive solution, the amount of hazardous waste destined for disposal was reduced from approximately 20,000 tons/year to 6000 tons/year. Efforts are under way to further reduce this amount because of EPA's land disposal ban.

In 1983, Ashland installed new process operations which included a state-of-the-art fluid bed, dry limestone pollution control system to remove SO_2 from the process flue gas. Operating year round, the system can generate in excess of 100,000 tons of spent limestone. The limestone used in the system is converted to calcium oxide (lime), which reacts with the SO_2 to form calcium sulfate. While we have a permitted industrial landfill for the calcium sulfate, a nonhazardous waste, we are constantly looking for ways to reuse this material, which still contains some reactive calcium oxide.

We have been able to find several uses for calcium sulfate, and we're always looking for more. Primarily we have used it as a chemical fixation, or sludge stabilization agent. By mixing it 1:1 with oily sludge we were able to eliminate a large old surface impoundment. Over a three-year period we chemically fixed approximately 90,000 tons of oily sludge. This stabilized material passed the EPA leachate test. We are looking at opportunities to market this material to others who have sludges requiring fixation or stabilization.

Ashland's coal company has also begun using the material for pH control of runoff from coal refuse disposal areas. The practice has been to add layers of lime every few feet of refuse fill to prevent the formation of sulfate bacteria and acid generation and leaching. The spent limestone was found to do just as good a job. We are also looking into its use in road construction. In many areas, small percentages of lime are mixed with clay to reduce plasticity and increase compressive strength. Testing has shown that our material may be a satisfactory substitute. Its use in the formulation of concrete is also being studied.

Because of the nationwide interest in waste minimization, the high cost of disposal, and the increased value of by-product recovery, more opportunities are becoming available, especially in the area of spent catalyst. We now regenerate certain types of spent catalyst. Other spent catalysts are sent to a metals recovery facility. We see, and use, more and more mineral by-product brokers, who for a fee and a percentage look for customers for waste materials and spent catalysts.

One often forgotten way to reduce waste is to solicit suggestions for waste minimization from the person on the line or in the plant. Many companies have for years provided incentives for employees to generate ideas to reduce costs and improve efficiency. Now there is an opportunity to extend that type of program into the waste minimization area. It only takes a few new slogans and management's promotion, and it doesn't cost much.

As was stated earlier, the amount of waste minimization accomplished at most facilities is only limited by the imagination of those concerned about waste minimization. Waste minimization may be a process to follow, but it is also an attitude—an attitude on the part of management and employees to make progress. In summary, we all know that waste minimization is here to stay. There is a great economic incentive today to reduce waste, and there will be an even greater incentive tomorrow. The cost of waste disposal will continue to go up, but even more importantly, the availability of waste disposal options will go down.

Ozone Depletion and Waste Minimization

Katy Wolf

Source Reduction Research Partnership
Los Angeles, California

18.1 INTRODUCTION

In 1974, two scientists hypothesized that chemicals called chloro-fluorocarbons (CFCs) were causing depletion of the ozone layer that protects the earth from harmful levels of ultraviolet radiation.[1] The fully halogenated CFCs,* so the theory went, were extremely stable. They did not break down in the lower atmosphere or troposphere but remained intact for decades. Eventually they made their way to the

---

*A fully halogenated CFC contains fluorine, chlorine, or bromine atoms, or combinations of them, exclusively. An HCFC is not fully halogenated if it contains a hydrogen atom.

upper atmosphere or stratosphere where ultraviolet light impinged upon them, causing them to decompose and liberate chlorine. The chlorine reacted catalytically with ozone, depleting the so-called stratospheric ozone layer and increasing the level of ultraviolet radiation reaching the surface of the earth. The increased radiation could cause a higher incidence of skin cancer and have a variety of deleterious effects on the ecosystems that support life.

In response to the threat of ozone depletion, the U.S. Environmental Protection Agency (EPA) and the Food and Drug Administration (FDA) banned the use of CFCs in aerosol applications in 1978.[2,3] At that time, such applications accounted for about half of the U.S. production of CFCs.[4] The EPA sponsored research on the impact of regulating CFCs and other ozone-depleting substances in nonaerosol applications and on the control measures for limiting emissions.[4,5] In the years that followed, it appeared that ozone depletion might be a less serious problem than was first anticipated,[6] and no further regulatory action was taken.

In 1985, the ozone hole in the Antarctic was detected. In the next few years, researchers tried to determine the cause of this hole, which was not predicted by the atmospheric models. Eventually there emerged convincing evidence that the hole was related to stratospheric ozone depletion and that the CFCs and other ozone-depleting substances were the cause. In the meantime, a series of international meetings culminated in an agreement in September of 1987 that would cap the production of the fully halogenated CFCs and halons* at 1986 levels and would reduce production of the CFCs by half over the next decade.[7] In December 1988, EPA proposed a regulation that would mimic the protocol domestically, and in August of 1989 the regulation was finalized.[8,9] The regulation would become effective in January 1989, and production of the ozone-depleting substances would be capped at 1986 levels beginning in July 1989.

With the advent of the domestic regulation, CFC producers are devoting impressive resources to research efforts for identifying and producing alternative CFCs. Indeed, DuPont—the world's largest CFC producer—has vowed to stop producing the ozone-depleting CFCs as soon as alternatives can be phased in.[10] CFC users are also trying to identify alternative chemicals, products, or processes that would minimize their use of CFCs until alternatives become available.

---

*Halons contain bromine, which is thought to pose a greater threat to the ozone than does chlorine.

18.2 PRODUCTION AND USE OF OZONE-DEPLETING SUBSTANCES

Table 18.1 presents the 1985 U.S. production level of the chemicals involved in ozone depletion, their major uses, and their ozone-depletion potential relative to that of CFC-11. The international agreement and the companion domestic regulation target CFC-11, -12, -113, -114, -115, and Halon 1301, 1211, and 2402. HCFC-22 and 1,1,1-trichloroethane (TCA) are excluded because they are not fully halogenated and accordingly have low ozone-depletion potential. Carbon tetrachloride, although it is fully halogenated, is excluded, because virtually all of it is used as an intermediate. It is therefore chemically converted to other substances and never emitted to the atmosphere. Ironically, its intermediate use is in the production of CFC-11 and CFC-12.

CFC-11 and CFC-12 are the two most widely used CFCs; indeed, the two are ubiquitous. CFC-11 is employed in the production of flexible foam, which is used in mattresses and carpet underlay, and in the manufacture of rigid foam, which is used as insulation in buildings. CFC-12 is used largely as a refrigerant in automobile air conditioning and in retail food stores. It is also employed widely as a blowing agent in packaging and insulating foams. HCFC-22 has a much lower ozone-depletion potential; it is used in home air conditioning and as an in-

TABLE 18.1   1985 U.S. Production, Use and Depletion Potential of Several Ozone-Depleting Substances

Substance	Production (thousand metric tons)	Use[a]	Ozone-depletion potential[b]
CFC-11	75.0	Blowing agent, refrigerant	1.0
CFC-12	135.0	Refrigerant, blowing agent	1.0
HCFC-22	123.0	Refrigerant, intermediate	0.05
CFC-113	73.2	Solvent	0.8
CFC-114	4	Intermediate	1.0
CFC-115	4	Refrigerant[c]	0.6
Halon 1301	5.4	Fire extinguishant	10.0
Halon 1211	2.7	Fire extinguishant	3.0
Halon 2402	d	Fire extinguishant	6.0
Carbon tetrachloride	280.0	Intermediate	1.1
TCA[e]	270.0	Solvent	0.1

[a] Substances are actually used in a range of applications. Indicated ones are major uses.

[b] Depletion potential as defined in the Federal Register notice proposing the domestic regulation. Depletion potential is relative to CFC-11, which has a defined value of 1.00.

[c] CFC-115 does not itself function as a refrigerant. It is combined with HCFC-22, and the blend, CFC-502, is a refrigerant.

[d] Production of Halon 2402 is negligible.

[e] Chemical name is 1,1,1-trichloroethane, which is otherwise known as methyl chloroform.

SOURCE: Refs. 9 and 11; author's estimates.

termediate in fluoropolymer production. Most CFC-114 is used to manufacture CFC-115, which is combined with HCFC-22 and used as a refrigerant in retail food stores. The Halons function as fire-extinguishing agents. Carbon tetrachloride, as mentioned earlier, is used to produce CFC-11 and CFC-12.

18.3   CFC-113

CFC-113—the chemical of focus here—is used primarily as a solvent. Its ozone-depletion potential is 80 percent of that of CFC-11. (TCA is also employed as a solvent, and, although its ozone-depletion potential is only 10 percent of that of CFC-11, its production level is very high).

18.3.1   Applications of CFC-113

Table 18.2 shows the various uses of CFC-113. Two-fifths of the chemical is used in the electronics industry, the majority of it for removing the flux from printed circuit boards after the components have been soldered to them. Critical cleaning accounts for 22 percent of CFC-113 use. It includes applications like cleaning disk drive assemblies and optical parts. Degreasing represents about one-fifth of CFC-113 use and involves grosser cleaning operations like degreasing pipes and other metal and glass parts. In other applications, CFC-113 is employed as a blowing agent in foams, as a refrigerant, and as a chemical intermediate.

18.3.2   Approach to waste minimization

CFC-113 will be regulated in the near future. In all likelihood, the producers will stop producing the chemical within the next decade.-

TABLE 18.2   CFC-113 Applications

Application	Use (thousand metric tons)	Percent
Electronics:	31	42
Defluxing	23	
Other*	8	
Critical cleaning	16	22
Degreasing	15	21
Other†	11	15
Total	73	100

*Other electronics applications include degreasing printed circuit boards and semiconductors.
†Includes use as refrigerant, a blowing agent, an intermediate, and a dry-cleaning agent.

Over that time frame, then, industry must phase out its use in virtually all applications. This phasing out suggests that identifying methods of reducing or eliminating CFC-113 will become increasingly important.

There are continuing debates about what constitutes waste minimization and source reduction. Although such debates may have a place in the political arena, they have little practical value in the day-to-day management of hazardous substances. It is clear from the preceding discussion that it is atmospheric emissions of CFC-113 that are of environmental concern. Because of the volatility of the chemical, however, even the CFC-113 waste that is generated eventually ends up in the atmosphere if it is not destroyed. In light of the fact that virtually all of the chemical will ultimately reach the atmosphere, the working definition of waste minimization in the balance of this chapter is a multimedia one, and it includes any action that minimizes or eliminates the use or release of CFC-113.

18.3.3 Waste minimization measures

The options for reducing or eliminating CFC-113 use and emissions fall into four generic categories: substitution, recycling and reuse, equipment modification, and improved housekeeping and operating practices. Substitution can be classified as chemical, process, or product substitution, i.e., other chemicals can be used; new processes that minimize the use of the CFC can be adopted, or existing processes can be modified; and products that do not use CFC-113 in their production process can be developed. Recycling can occur through capture and reuse of the CFC atmospheric emissions; it can also occur through reclamation of contaminated solvent waste. Existing solvent equipment can be modified to reduce the level of atmospheric emissions. Better operating practices can reduce the use and release of CFC-113. Each of these options is discussed below.

Chemical substitution. There is a whole range of substances that might be appropriate to replace CFC-113 in the applications listed in Table 18.2. In general, conversion to these other substances carries with it trade-offs of one sort or another. The alternatives may themselves be dangerous but simply in a different way; they may convert the problem to a new medium; or they may be more costly to use. In general, the decision of whether or not a chemical is a reasonable alternative must be made on a case-by-case basis. Nevertheless, I try to make some general comments on the applicability for the end uses listed in Table 18.2. A few of the most important chemical substitutes are mentioned in what follows.

Other chlorinated solvents. The four major chlorinated solvents, trichloroethylene (TCE), 1,1,1-trichloroethane (TCA), methylene chloride (METH), and perchloroethylene (PERC), are shown in Table 18.3, together with some of their characteristics; CFC-113 is included for comparison. The chlorinated solvents are most appropriate as alternatives to CFC-113 in the degreasing applications specified in Table 18.2. They might also be applicable in the "Other" electronics applications called out in Table 18.2 and in critical cleaning. TCA is used to some extent today in "defluxing" applications.

Two of the solvents—TCE and PERC—have low allowable workplace exposure levels. In contrast, because it is nontoxic, CFC-113 has the highest workplace exposure level allowed. TCE, once the most widely used solvent, has largely been replaced by TCA. TCE is photochemically reactive, and, like PERC, is not exempted as a volatile organic compound (VOC) under Section 111 of the Clean Air Act. TCA and METH (and also CFC-113) are exempted. Although the results remain controversial, three of the solvents—TCE, PERC, and METH—have given positive results in animal carcinogenicity tests. EPA has published an intent to list these three solvents as hazardous air pollutants under Section 112 of the Clean Air Act. TCA is presently undergoing an animal carcinogenicity test. This chemical, as indicated in Table 18.1, does cause ozone depletion, but to a much smaller extent than CFC-113.

TABLE 18.3   Characteristics of Chlorinated Solvents

Solvent	PEL*	Smog	Potential carcinogen	Ozone depleter	Hazardous air pollutant
TCE	100(25)	Yes	Yes	No	Intent†
PERC	100(5)	Yes	Yes	No	Intent
METH	500†	No	Yes	No	Intent
TCA	350	No	In progress	Yes	—
CFC-113	1000	No	No	Yes	—

*PEL is the time weighted average. It is the allowable exposure level in the workplace for an 8-hour workday (40-hour workweek). In general, the lower the PEL, the more dangerous is the chemical. Values are set by the Occupational Safety and Health Administration (OSHA). Values in parenthesis are new proposed OSHA levels.

OSHA will propose a much lower PEL for METH in the future.

†Intent to list.

SOURCE: Ref. 12 and 13.

Many users are moving away from the chlorinated solvents TCE, PERC, and METH because of concern for workers and because local air districts regulate VOCs stringently. Some of these users have adopted TCA and CFC-113 in recent years. Now, with the impending regulation on CFC-113, TCA is the only remaining option. However, TCA may itself eventually be regulated as an ozone depleter.

CFC-113 is a very mild solvent with low solvency power. The other four chlorinated solvents are harsher cleaners than CFC-113. In many instances, CFC-113 has been deliberately chosen for a given application for its gentle cleaning action, and the other solvents may not be appropriate substitutes. For instance, in printed circuit board defluxing, CFC-113 solvents are chosen in part because they are compatible with the substrate materials that make up the board.[14] TCA is used somewhat for defluxing applications, but its applicability is limited by its aggressiveness toward the substrate materials.

Because of the existing regulatory regime, the potential for substituting the other chlorinated solvents for CFC-113 is low. Although TCA is under less regulatory scrutiny than the other chlorinated solvent presently, it may be regulated in the future, and therefore offers limited promise as an alternative.

Flammable solvents. These solvents were used for many years but were considered dangerous because of the threat they pose to the worker. They were replaced in many cases by the chlorinated solvents, which are not flammable and were considered safer alternatives at the time. Some of the flammable solvents—including hydrocarbons, alcohols, and ketones—might be appropriate substitutes for CFC-113. Acetone and perhaps a few of the other flammable solvents might replace CFC-113 in electronics applications such as degreasing semiconductors. Flammable solvents may also offer some potential in CFC-113 degreasing applications.

Pure alcohol is an excellent defluxing agent.[14] Its drawback is its flammability, which poses a threat to workers, leads to high insurance costs, and increases the costs for making a facility fire-resistant. Another drawback of alcohol, and indeed of the other flammable solvents as well, is that they are photochemically reactive and are not exempt from Clean Air Act regulations. Thus, in many localities where VOC limits are imposed, users would not be able to convert to them.

Combustible solvents. A new class of combustible solvents has appeared recently on the market. These include N-Methyl-2-pyrrolidone (NMP) and dibasic esters (DBE), which are used primarily as paint-stripping agents, and terpenes, a citrus extract. Since these solvents are not exempted, all are VOCs and are restricted under the Clean Air Act. However, it is also true that all three of these chemicals have low vapor pressures, and it is unlikely they would be emitted in large quantities as they are used. Each of the solvents is combustible, and it is not yet clear what the related fire insurance requirements will be.

Terpenes have proved to be very good defluxing agents,[15] but they have not been listed as allowed solvents for this application under military specifications.[16] An effort is under way to change these specifications to allow terpene use. NMP, DBE, and terpenes all might be used as alternatives to CFC-113 in the degreasing and the critical cleaning applications in Table 18.2. A drawback of NMP is its low PEL (see Table 18.3) of 100 ppm. A drawback of terpenes is that the major ingredient—D-limonene—has given a positive carcinogenicity test in male rats.[17]

CFC blends.   Today, CFC-113 is commonly combined with alcohol in some applications. The chemical forms an azeotrope (a blend of two or more chemicals that boils at a constant temperature) at small concentrations of alcohol. CFC-113 use could be reduced by increasing the amount of alcohol as the second component. Disadvantages are that the alcohols are not exempted as smog contributors and that nonazeotropic blends can become alcohol-rich, and therefore flammable.

Other blends are being investigated as well. One is a blend of CFC-113 with 1,1-*trans*-dichloroethylene, a VOC. Another is a combination of CFC-113 with TCA and alcohol. These blends hold only limited promise for stretching out CFC-113 use in the next decade.

New HCFCs.   The worldwide CFC producers are investigating new HCFCs that might serve as alternatives to CFC-113. These generally contain hydrogen so they won't pose as great a threat to the ozone layer. One alternative suggested several years ago—HCFC-132b—later proved to be toxic, and work on it was discontinued. Research on other possible candidates is continuing.

Process substitution.   The major process alternative to CFC-113 cleaning is aqueous cleaning. Other options include low- or no-clean flux and use of fluxless solder. These alternatives are discussed below.

Aqueous cleaning.   This process is most applicable for the defluxing and degreasing applications specified in Table 18.2. Although military specifications for defluxing allow the use of water, water-dispersible fluxes are not permitted. This effectively means that water cannot be used. In critical cleaning and other electronics applications (i.e., not defluxing), water may prove promising in some cases.

Water is already commonly used for defluxing printed circuit (PC) boards.[14,18] There is a movement in PC board fabrication toward so-called surface mount technology (SMT). Components are mounted on the surface of the boards, and the trend to further miniaturization means that there is a very low clearance—perhaps 4 mm spacing—between the boards and the components. After the components are soldered to the board, a flux, which must ultimately be removed, is applied to the leaded components. At very low clearances, the contact angle can prevent water from removing the flux from under components. Other solvents, like CFC-113, do not have this problem. Another drawback of water is that local regulations may prevent its direct release to the sewer. Metal concentrations after defluxing may be high.

The effort to change military specifications may allow the increased use of water for defluxing applications. Other users are investigating aqueous cleaning for degreasing[19] and for critical cleaning applications. In that light, one large international firm is developing a process for using water to dry disk head assemblies.

In all applications, converting to water requires purchase of new equipment and, in some cases, modification of assembly line processes as well. Water does not dry as quickly as the volatile CFC-113, so a longer drying time or drying equipment like a hot air knife may be necessary.[14]

Low- or no-clean flux. This option applies only to the PC board defluxing application. There are low-solids fluxes on the market that are used when the PC boards are destined for noncritical uses, as in toy manufacture. The flux is simply left on the board and allowed to remain as the board is placed in the item. No-clean fluxes are also being developed for applications where cleanliness is essential. These low-solids fluxes would eliminate the use of CFC-113.

Product substitution. The concept of molecular electronics, in which organic molecules themselves act as semiconductors, is being investigated at Carnegie Mellon. This development, which will not be demonstrated for many years, would make degreasing of semiconductors obsolete.

Recycling of vapors. The major method for capturing CFC vapor emissions is carbon adsorption. The most common arrangement is to pull the solvent vapors and air through the lip vent of a degreaser and route them to a carbon adsorber which contains activated carbon.[5] The CFC-113 is adsorbed to one of the carbon beds. The second carbon bed can adsorb solvent while the first bed is desorbed, most commonly with steam. The desorbed solvent and water are separated physically or by distillation, and the CFC-113 can be reused in the degreaser.

There are several disadvantages to this process. First, carbon adsorption systems are expensive. In many parts of the country, such as in California, for instance, few users have installed steam. The requirement for on-site steam increases the cost of the system even more. Second, carbon adsorption is more difficult to use with blends or azeotropes of CFC-113. The second component is frequently an alcohol or acetone, and these substances are soluble in water. During steam desorption, these components will remain in the water phase, and the CFC-113 must be reblended before reuse. Third, there is some expertise required to operate the system, which many users do not have. Fourth, carbon adsorption can create water pollution problems if the solvent and water are not adequately separated.

Recycling of waste. Some users recycle their contaminated liquid solvent on site for reuse in the original process. Stills are commonly employed either in a batch or continuous mode for this purpose. On-site recycling can reduce the virgin solvent requirement; it also reduces total releases because less CFC-113 is produced.

There are certain disadvantages of on-site recycling. First, if a nonazeotropic blend is used, the components will be separated in the distillation, and the user will be required to reblend before reuse. Second, many users employ so-called pot stills, and the residue frequently still contains a significant fraction—as much as 35 percent—solvent. If the user has expertise, which many do not, this level can be reduced substantially. Third, the residue remaining after distillation, the still bottom, requires disposal as hazardous waste. The land disposal ban on solvents implies that the sludge must be incinerated, which can be extremely costly.

Off-site recyclers will pay users for CFC-113 solvents if they are not highly contaminated. Recyclers reclaim the solvent with distillation or thin-film evaporation and sell it back either to the same user or to a new customer. Advantages of off-site recycling are that the recycler deals with the reblending and sludge disposal. There are three main disadvantages of off-site recycling (which, incidentally, are advantages for on-site recycling). First, the waste has to be transported, possibly endangering communities. Second, the reconstituted solvent contains tiny amounts of contaminants that arise in other processes and may cause quality control problems. Third, the liability remains with the generator if the recycling firm is incompetent.

In California, recyclers and users have established an interesting procedure for dealing with chlorinated solvents. There are no destructive incinerators in California, but there is a cement kiln that accepts hazardous waste. Recyclers collect contaminated solvent from users, reclaim it, blend the sludge with flammable solvents to increase the heating value, and send it to the cement kiln as supplementary fuel. Recyclers will also accept sludge from users who practice on-site recycling if the sludge still contains a large amount of solvent. Users who have an efficient on-site recycling process are not able to dispose of their sludge in the cement kiln, because the kiln does not accept small volumes of waste from users. The user must send the sludge out of state for destructive incineration at a higher cost. This system discourages efficient on-site recycling and encourages off-site recycling where the recycler acts as the middleman.

Microfiltration—another form of recycling—can be used in a degreaser to remove contaminants continuously, stretching out the useful life of the solvent. Such systems remove particles less than 0.1 micrometers in size, undissolved water, acids, and organic and inorganic soils.

Equipment modifications. Equipment for use with CFC-113 has generally been conservative of solvent losses because of the expense of CFC-113-based solvents as compared to other solvents. Nevertheless, for users with older equipment, there are many equipment changes that can reduce solvent loss.

The freeboard is the portion of the degreaser above the liquid zone. Increasing the freeboard ratio to 1 can minimize losses. This ratio is defined as the freeboard height divided by the width of the degreaser. Fitting the degreaser with a refrigerated chiller is a good method for controlling vapor losses. The system employs condenser coils and a freeboard water jacket. If the operation lends itself to conveyorized equipment, solvent losses can be significantly reduced. In contrast to open-top degreasers, conveyorized equipment is generally enclosed, except for entrance and exit ports for the parts to enter and leave.

Better operating practices. A host of improved operating practices can effectively minimize solvent losses.[20] Such measures include leaving the part in the degreaser until the solvent dripping ceases, covering equipment when it's not in use, and avoiding the use of spray above the vapor level.

18.4   SUMMARY AND CONCLUSIONS

There are a variety of options for reducing the use and release of CFC-113, one of the commonly used chemicals that depletes the ozone layer. Current agreements and regulations will reduce production of the chemical to one-half its 1986 production level within the next decade. The largest CFC producer has vowed to phase out production altogether as substitutes are identified. Over the next ten years, some users will adopt substitutes and some will continue to use CFC-113 more conservatively until new alternatives are developed.

The waste minimization measures discussed here hold promise for reducing and eliminating CFC-113 use. Table 18.4 summarizes the options qualitatively and evaluates them over two dimensions. The first dimension is the potential of the option in reducing CFC-113 use. The second dimension is the time frame for implementation.

Because of the possibility of using TCA, the potential reduction in CFC-113 by substitution is medium. Because TCA is available today and can be used in similar equipment, the substitution could be accomplished immediately. Flammable solvents are less promising because they are VOCs. Combustible solvents could have more impact, but more time is required to develop compatible equipment. CFC-113 blends have only small potential. New HCFCs could effectively replace CFC-113 altogether, but they wouldn't be available for many years because of toxicity testing.

Water offers a fair amount of promise over the medium term as firms perform research and development (R&D) on the best way to use aqueous cleaners for their processes. Low-solids fluxes offer significant promise. Molecular electronics may eventually make solvents obsolete, but only over the very long term. Recycling of vapors has limited applicability, although the systems are available today. Recycling of waste is a good option that could be implemented immediately. Because CFC-113 equipment is generally already fairly conservative, equipment modification will have only a modest effect. Operating practice improvements will have a similar impact.

TABLE 18.4   Options for Reducing CFC-113 Use

Option	Potential reduction	Time frame*
Other chlorinated solvents	Medium	Short-term
Flammable solvents	Small	Short-term
Combustible solvents	Medium	Medium-term
CFC-113 blends	Small	Short-term
New HCFC solvents	Large	Long-term
Aqueous processes	Medium	Medium-term
Low-solids flux	Medium	Medium-term
Product substitution	Large	Long-term
Recycling of vapors	Small	Short-term
Recycling of waste	Medium	Short-term
Equipment modification	Small	Short-term
Improved operating practices	Small	Short-term

*Short-term implies immediate; medium-term is 2 to 5 years; long-term is more than 5 years.

In the next several years, as regulations go into effect, most users will adopt the short-term measures of Table 18.4 that minimize their use of CFC-113. They will convert to TCA and other CFC blends; they will recycle their waste; and they will modify their equipment and improve their operating practices. Over the medium- and long-term, as CFC-113 is phased out, these users will convert to combustible solvents, aqueous processes, low-solids fluxes, and new HCFC solvents as they become available. There is no one chemical substitute for CFC-113 in all its applications, and it is not likely that there ever will be. Alternatives will be found and adopted on a case-by-case basis in the years to come.

18.5 REFERENCES

1. M. J. Molina and F. S. Roland, "Stratospheric Sink for Chlorofluoromethanes: Chlorine Atom-Catalyst Destruction of Ozone," *Nature*, 249, p. 810.
2. *Federal Register*, 43, March 17, 1978, p. 11301.
3. *Federal Register*, 43, March 17, 1978, p. 11318.
4. A. R. Palmer, W. E. Mooz, T. H. Quinn, and K. A. Wolf, "Economic Implications of Regulating Chlorofluorocarbon Emissions from Nonaerosol Applications," The RAND Corp, R-2524-EPA, June 1980.
5. W. E. Mooz, S. H. Dole, D. L. Jaquette, W. H. Krase, P. F. Morrison, S. L. Salem. R. G. Salter, and K.A. Wolf, "Technical Options for Reducing Chlorofluorocarbon Emissions," The RAND Corp, R-2879-EPA, March 1982.
6. National Academy of Sciences, "Causes and Effects of Changes in Stratospheric Ozone: Update 1983," National Academy Press, Washington, D.C., 1984.
7. *Montreal Protocol on Substances That Deplete the Ozone Layer*, Final Act, Montreal, Canada, September 1987.
8. *Federal Register*, 52, December 14, 1987, p. 47485.
9. *Federal Register*, August 12, 1988, p. 30566.
10. DuPont Update, "Fluorocarbon/Ozone, New DuPont Position Stresses Orderly Transition to Alternatives," July 1988.
11. J. K. Hammitt, K. A. Wolf, F. Camm, W. E. Mooz, T. H. Quinn, and A. Bamezai, "Product Uses and Market Trends for Potential Ozone Depleting Substances, 1985–2000," The RAND Corp, R-3386-EPA, May 1986.
12. *Federal Register*, 54, January 19, 1989, p. 2332.
13. K. Wolf and F. Camm, "Policies for Chlorinated Solvent Waste—An Exploratory Application of a Model of Chemical Life Cycles and Interactions," The RAND Corp, R-3506-JMO/RC.
14. P. Morrison and K. Wolf, "Substitution Analysis: A Case Study of Solvents," *J. Hazardous Materials*, vol. 10, May 1985.
15. AT&T News Release, "AT&T Plays Major Role in Developing Substitute for Compound Linked to Depletion of Ozone," January 13, 1988.
16. Military Standard, "Soldering Technology, High Quality/High Reliability," DOD-STD-2000-1B, January 15, 1987.
17. *NTP Technical Report on the Toxicology and Carcinogenesis Studies of d-Limonene in F344/N Rats and B6C3F1 Mice*, final report expected in January 1989.
18. SMT, "High Pressure Spray and Ultrasonic Cleaning for Surface Mounted Assemblies," Surface Mount Technology, December 1987, p. 11.
19. S. Evanoff, K. Singer, and H. Weltman, "Alternatives to Chlorinated Solvent Degreasing—Testing, Evaluation and Process Design," General Dynamics, in *Process Technology 88*.
20. U.S. Environmental Protection Agency, "Control of Volatile Organic Emissions from Solvent Metal Cleaning," November 1977.

Chapter 10

Case Studies of Successful Solvent Waste Reduction

Robert H. Salvesen

Robert H. Salvesen Associates
Tinton Falls, New Jersey

19.1   Background

This section discusses successful solvent waste reduction practices that utilize on-site or off-site distillation equipment. While solvent reclamation by distillation has been practiced for many years in a wide variety of industries, this practice has expanded considerably among both large and small businesses. This expansion has been brought about because of environmental regulations, but most solvent users have also found recycling to be economically attractive.

The question of whether a solvent user should recycle on site or off site depends upon a number of factors, such as location, volumes and types of solvents, economics, and management preferences. A good environmental audit and economic analysis are required to determine the most cost-effective option for a specific facility.

Solvent recycling is used by laboratories, hospitals, asphalt manufacturers, small shops (electric motor rebuilders), and all types of industries, including giants such as Du Pont and IBM. Low-boiling, pure solvents are the easiest to recycle, but a wide range of solvents boiling up to about 400°F are commonly reclaimed. A list of the various types of solvents most frequently reclaimed is provided in Table 19.1. Table 19.2 gives a list of industries and manufactured products in which solvents are used and recycled. In the future, these lists will continue to expand to include other solvents and industries because of both environmental and economic benefits.

There are a growing number of equipment suppliers who can provide on-site distillation units which are simple and easy to operate. The most common unit sizes, in gallons, are 3, 5, 15, 55, 250, and 500. Many manufacturers offer a number of features, including vacuum stills. Smaller units require 20 to 30 minutes of operator time per batch (6- to 8-hour shifts), and larger units may require 1 to 2 hours

TABLE 19.1   Solvents Most Commonly Reclaimed

Category	Solvent
Hydrocarbons	VM&P naphtha Mineral spirits Aromatic naphtha Toluene Xylene Isoparaffinic Turpentine
Glycol ether	Cellosolves
Alcohols	Methyl Ethyl Isopropyl
Ketones	MEK MIBK
Esters	Ethyl acetate Butyl acetate
Mixtures containing solvents listed opposite	Methylene chloride Mineral spirits Toluene/xylene Ketones Esters Alcohols Phenols

TABLE 19.2   Industries and Manufactured Products Which Utilize Recyclable Solvents

Product	Solvents
Labels	Ketone
Seals and gaskets	Chlorinated blend
Aluminum castings	Chlorinated/alcohol
Electronic components	Chlorinated solvents
Plastics decoration	Acetone/MEK
Chemicals	Chlorinated solvents
Packaging	Alcohols
Buttons	Solvent mix
Printing	Toluene
Molded plastics	MEK/toluene
Wire	Toluene
Adhesives	Toluene
Pharmaceuticals	Naphtha
Ovens	Toluene
Graphite	Toluene, acetone
Electronics	Chlorinated solvents
Metal fabrication	Chlorinated solvents
Car interior parts	MEK
Shoes	Ketones
Screens	Chlorinated solvents
Quartz crystals	Freon/acetone

per batch. Illustrations of several units are provided in Figures 19.1 to 19.4.

Selection of the best unit for a facility or application should be made after careful consideration of the needs and evaluation of the various features available from equipment suppliers. Units are commonly placed near the source of used solvent generation, but centralized systems servicing a number of generators have also proved satisfactory. Since these on-site stills are simple, they will not separate mixtures of solvents. Thus, good segregation is essential, and solvent stills must be either designated for one type of solvent or cleaned between batchs of different solvents. Commercial recyclers or custom-built stills are capable of separating solvent mixtures, but these operations require highly trained operators. Experience has shown that the quality of re-

Figure 19.1   A 55-gal/day atmospheric still. (*Photograph courtesy of the Finish Company, Inc., Greengarden Road, Erie, Pa.*)

cycled solvent is generally equal to or better than new-product specifications. If necessary, filters and water removal equipment can be added to improve product quality.

19.2   Solvent Recycling: Robins Air Force Base

Robins Air Force Base has saved about $600,000 per year by recycling a variety of solvents, oils, and fuels, utilizing different types of equipment.[1] This base is one of the many Department of Defense (DOD) facilities that have responded to the mandates issued by DOD to minimize waste generation and to maximize recycling and reclamation of all types of used materials. In most instances the major wastes are used oils and solvents produced by the many industrial operations required in maintenance and overhaul of DOD equipment.

Units have been installed at a centrally located facility on the base. The materials and annual volumes being recycled are noted below:

Freon 113: 2200 gal/year

Figure 19.2   Cutaway of a 760 gal/day still. (*Photograph courtesy of the Finish Company, Inc., Greengarden Road, Erie, Pa.*)

Heat transfer fluid: 1400 gal/year

Degreasing solvents: 13,000 gal/year

Isopropyl alcohol: 4200 gal/year

Purging fluid: 30,000 gal/year

JP-4 fuel: 3000 gal/year

Silicone damping fluid: 3400 gal/year

Dry-cleaning solvent: 2400 gal/year

Coolanol 25R fluid: 250 gal/year

Equipment utilized to reclaim the above materials included solvent stills and filtration units. The recycled material met original product specifications.

Figure 19.3 A 55 gal/day still with top loading and a bottom cleanout opening. (*Photograph courtesy of Progressive Recovery, Inc., 1020 North Main Street, Columbia, Ill.*)

19.3   Solvents: Norfolk Navy Shipyard

Paint solvents, Freon PCA, and mineral spirits have been recycled at the Norfolk Naval shipyard for a number of years, resulting in savings of over $50,000 per year.[2] For recycling Freon, a thin-film evaporator has been employed for over 20 years and has produced a high-purity product meeting the very strict specifications for low residue and particulate requirements. To reclaim the mineral spirits, a vacuum still was needed to recover this high-boiling hydrocarbon solvent. The vacuum still was required to keep the temperature below the autoignition temperature of the solvent. An atmospheric still was adequate for recovery of the more volatile paint thinners.

Figure 19.4 A scraped surface distillation unit used for high-solids or high-viscosity materials. (*Photograph courtesy of Progressive Recovery, Inc., 1020 North Main Street, Columbia, Ill.*)

19.4 Freon 113: Naval Shipyard

At another large Naval shipyard, about 30,000 gal/year of Freon 113 and PCA are recycled using a simple pot still.[3] Since this type of still is not as efficient as the thin-film evaporator noted in the previous example, it was necessary to follow the distillation with filtration and silica gel treatment to remove particulates and water, respectively. Annual savings amounted to over $300,000 per year.

19.5 Small Distillation Units: Department of Defense (DOD) Facility

Fifteen separate distillation units have been recommended at a major DOD aircraft repair facility in California.[3] The decision was made to recommend many small units rather than a few large units to enable the various shops to maintain close control over quality control. In addition, the costs of transportation would have added considerably to the total reclamation costs.

These batch stills are to be placed in major shops generating used solvents. The types of solvents generated and annual volumes are noted below:

Dry-cleaning solvent (mineral spirits): 22,000 gal/year

Ethyl acetate: 4800 gal/year

1,1,1-trichloroethane (1,1,1-TCA): 14,000 gal/year

Paint thinner: 18,000 gal/year

Freon 113 and PCA: 9000 gal/year

Total savings are anticipated to be over $500,000 per year. A contributing factor to these savings is the high cost of disposal at this location.

19.6   On-Site Distillation: DOD Facility

By installation of on-site distillation equipment at a large overseas DOD facility, savings of over $200,000 per year are anticipated.[3] In addition to over 5000 gal/year each of Freon 113, 1,1,1-TCA, and mineral spirits, about 2500 gal/year of a special water-based cleaning solution was generated. These used materials were all reclaimed by distillation. Of the total $200,000 per year saved, about $50,000 per year was saved by distilling off water from the aqueous cleaning solution. Disposal costs at this remote facility were almost $20 per gallon, so the large reduction in volume made this treatment very attractive economically.

19.7   Batch Distillation: Asphalt Manufacturer

A small road asphalt manufacturer in New Jersey has installed a 5 gal/day batch distillation unit to recycle used 1,1,1-TCA generated in cleaning laboratory test equipment.[3] While this cost about $5000 to install, it is anticipated that the payback period will be less than 2 years. The manufacturer believes that installation of the distillation unit will protect the company from potential liability caused by alternate disposal options. The quality of product recovered was excellent.

19.8   Solvent Recycling: Hospital

Hospitals are saving from $10,000 to $20,000 per year by recycling solvents used in histology laboratories.[4] High-purity ethanol, toluene, and xylene are recycled in specially constructed spinning band distillation units capable of producing products that meet stringent hospital standards. These units cost from \$5,000 to \$10,000, or more, and are generally small enough (5- to 10-gal capacity) to be placed in a standard laboratory hood.

19.9 TCA Recovery: Electric Motor Reconditioner

A small reconditioner of electric motors, controls, and fuel pumps generates about 15 gal/day of used 1,1,1-TCA.[3] Using a small 15 gal/day distillation unit, clean solvent is recovered which is satisfactory for reuse without additional treatment. A person working on the equipment cleaning operation devotes about 20 minutes each day to tending the recycling operation. The payback period has been estimated to be less than two years. The type of reclamation unit used is shown in Fig. 19.1.

19.10 Solvent Recovery: Printing Equipment

Solvents used to clean printing equipment and to thin inks are reclaimed by use of simple distillation units.[3] For printing on polyethylene or polypropylene, the solvents employed are mixtures of ethanol and n-propyl acetate. These materials form an azeotrope at the distillation temperatures which corresponds exactly to the desired solvent mixture. No testing or quality control is required for the reclaimed solvents. This user indicates a payback period of from 1 to 2 years for the two solvent recovery units.

19.11 Solvent Recycling: Electronic Equipment

A major manufacturer of electronic equipment, computers, etc., has installed a number of solvent-recycling units throughout its operating facilities.[3] The major solvents reclaimed are Freon 113 and mixtures of Freon with isopropyl alcohol. Rather than using centralized facilities, the manufacturer decided to place the recycling units near sources of used solvent generation. Since there is a need for high-purity solvents, filters for particulate removal and alumina drying columns have been installed where necessary. Product quality is carefully checked, and most reclaimed solvent has met stringent specifications for low residue and particulate content. Payback periods for capital investments have been found to vary from 6 to 24 months.

19.12   References

1. Personal communication with O. H. Carstarphen of Robins Air Force Base, Georgia.
2. From an article about solvent reclamation in the *Norfolk Pilot*, November 5, 1984.
3. Personal communications.
4. *The International Hospital Federation Official Yearbook 1986*, Sabrecrown Publishing, P.O. Box 839, Garfield House, 86–88 Edgware Road, London W22YW.

Index